The Women Who Popularized Geology in the 19th Century

Kristine Larsen

The Women Who Popularized Geology in the 19th Century

Springer

Kristine Larsen
Geological Sciences
Central Connecticut State University
New Britain, CT, USA

ISBN 978-3-319-64951-1 ISBN 978-3-319-64952-8 (eBook)
DOI 10.1007/978-3-319-64952-8

Library of Congress Control Number: 2017950127

Printed on acid-free paper

This Springer imprint is published by Springer Nature
The registered company is Springer International Publishing AG
The registered company address is: Gewerbestrasse 11, 6330 Cham, Switzerland

Preface

Historians of science acknowledge that the late eighteenth through mid-nineteenth centuries represent a pivotal era in the evolution of the geological sciences, in which geology became a bona fide and well-established scientific endeavor. At the same time, science was increasingly brought into the public eye, largely thanks to public demonstrations and lectures aimed at nonspecialists, at both professional institutions (such as the Royal Institution in London) and more informal ones (including the Lyceum movement in the USA). The availability of what would now be termed *free-choice learning opportunities* at the popular level opened up science to a significant new audience, women as well as men without formal education in the sciences. The education of children was also undergoing substantive changes, including the formation of girls' schools (at both the elementary and secondary levels), providing parents with options beyond private tutors and home schooling. But teachers, tutors, and homeschooling mothers needed to learn both the science of the day and effective pedagogies that could be used to make science both understandable and appealing to children.

This was also the heyday of the literary style known as the *familiar format*, books written largely *by* women *for* women and children in the form of letters, conversations, and catechisms. These works were commonly set in a domestic setting and featured children (both boys and girls) actively exploring the natural world through hands-on activities, experiments, and direct observations, guided by a patient and knowledgeable mother figure. This style of science popularization played a central role in both formal and informal science education during the same time period that geology was becoming a well-defined science. It is therefore surprising that a study of women's popular-level geological writing during this period has not yet been published. This work endeavors to fill some of that gap in our understanding of the role of woman popularizers, specifically those utilizing the familiar format, in shaping public understanding of geology in the first half of the nineteenth century.

Chapter 1 begins by surveying the status of geology in the early nineteenth century, including debates over terminology, physical processes, theology, and the accepted role of women. After introducing some of the longstanding political issues with the popularization of science in general, the bulk of Chap. 2 focuses on the birth of the familiar format and illustrates the wide diversity of works that fall under its umbrella. Chapters 3–8 form the core of this volume, analyzing the lives, literature, and resulting lessons of six women who authored familiar format geological science books—Jane Marcet, Delvalle Lowry, and Maria Hack (Europeans) and Jane Kilby Welsh, Delia Woodruff Godding, and Almira Hart Lincoln Phelps (Americans). While these authors often found some level of fame in their lifetime, they have largely been forgotten today. This work is a first step toward righting that wrong.

Due attention will then be paid in Chap. 9 to three examples of women popularizers of geosciences during this era who elected not to write in the familiar format—Mary Roberts, Rosina Zornlin, and Mary Somerville—situating them in contrast to their colleagues. Finally, Chap. 10 explores the societal factors that led the familiar format to, metaphorically, go the way of the dinosaur, through the lives and writings of three women who, in their own ways, negotiated this transition to a more modern (and presumed to be more masculine) style of popularizing science: Arabella Buckley, Agnes Giberne, and Lady Grace Anne Prestwich.

This is the story of uncommon women who persevered in the face of a society that severely curtailed their access to formal education and male critics and rivals who alternately praised and ridiculed their popular level writing. These women persevered despite personal adversity, financial uncertainty, and onerous familial responsibilities, and at a time when the very science they were attempting to encapsulate was in the midst of seismic shifts of its own. They breathed life into the sometimes stiff and staid story of rocks and geological strata, and not only educated generations of children, but encouraged woman and men, young and old, beginner and practitioner, to think more deeply about the planet on which we all live. This book is an attempt to repay a portion of the debt that prior generations of children, and adults, owe to these women who persevered.

New Britain, CT Kristine Larsen

Contents

Chapter 1
The Status of the Geological Sciences Circa 1800

1.1 Debating Definitions

In her 1852 volume *Outlines of Geology for Families and Schools*, Rosina Zornlin explains that "Geology consists of an inquiry into the structure of the earth, and the nature and arrangement of the materials of which it is composed" (Zornlin 1852: 1). Such a definition would not seem out of place in a similar volume today, with the caveat that it does not mention the processes that shape the earth. However, it is important to note that geology as a science had undergone an, at times, particularly difficult and painful birthing process between 1780 and 1830, one rife with tensions between science and religion as well as sometimes vociferous political debates between different theoretical schools (Laudan 1987). It is also important to understand that the fundamental vocabulary by which geology and other sciences are described today was not universally agreed upon in the first half of the nineteenth century. For example, before 1900 the term *natural history* roughly covered what today would include the biological and geological sciences, while *natural philosophy* encompassed the topics of physics and chemistry. It is also essential to recognize that not only was geological knowledge limited at the turn of the nineteenth century, but also the general understanding of chemistry and physics. For example, less than 40 chemical elements had been observed in nature by 1800. Dalton's atomic theory of matter debuted in the first decade of that century, but Maxwell's unified theory of electromagnetism would have to wait another five decades to be codified, and radioactivity only entered the scientific imagination in the last decade of the nineteenth century.

The term *mineral*, which today refers very specifically to the naturally found chemical compounds that form the components of rocks, was once used far more broadly, as one of the three kingdoms of nature (animal, mineral, vegetable). Therefore in the late eighteenth and early nineteenth centuries *mineralogy* was the general umbrella term under which one could find not only the study of modern minerals, but rocks and fossils as well. Mineralogy was also, by necessity, intimately

K. Larsen, *The Women Who Popularized Geology in the 19th Century*,
DOI 10.1007/978-3-319-64952-8_1

connected to chemistry, and one could find a great deal of chemical knowledge contained within mineralogy texts. By the 1790s specific subdisciplines with a narrower focus emerged in the writings of Abraham Gottlob Werner (1749–1817) and were commonly used by fellow German scientists for the next three decades. These included *oryctognosy* (the identification and classification of minerals), *mineral geography* (the geographical distribution of rocks and minerals), *economic mineralogy* (the uses of minerals by and benefits to human society), *chemical mineralogy* (the chemical analysis and composition of minerals), and *geognosy* (the description of rock formations and their relationships to each other) (Guntau 2009). This last term derives from the invented term *geognosie*, or *earth-knowledge*.

All of these were very practical, observation-based modes of science, and largely avoided theoretical considerations. There was also (as previously noted), a decided lack of attention to geological processes in this classification scheme. These terms were also not universally utilized, especially beyond Germany. Such controversies concerning terminology often made their way into the popular-level writings and book reviews of the day. For example, in her *Conversations on Mineralogy* (1822), Delvalle Lowry's young character Frances asks about the difference between mineralogy and geology. Her tutor, Mrs. L, explains

> Mineralogy, in its most extensive signification, is the science of the unorganized mass of the globe; in which sense, it is divided into *Oryctognosy*, or a knowledge of minerals by their external characters; *Chemical mineralogy*, which teaches us to distinguish them by a knowledge of their composition; and *Geology* (or Geognosy), which makes us acquainted with the different relations of the mineral masses, which compose what is termed the *crust of the earth*. (Lowry 1822: 15–6; emphasis original)

She later notes that Oryctognosy is described as "the alphabet of geology" (Lowry 1822: 16). In contrast, a review of the book in *The Quarterly Journal* (1823, 155) criticizes this terminology, offering that "There is more parade than profit in this distinction. Mineralogy and geology are quite sufficient without oryctognosy—a lately introduced and bad term." Similarly, in her 1833 work *Familiar Lessons in Mineralogy and Geology*, Jane Kilby Welsh's Mr. G explains to his family that "'The term Geology… is significant of the science of the earth, and is derived from the Greek for *earth* and *reason*. Werner, and some of the French authors, term it Geognosy; but the former appellation is the most generally received at the present time" (Welsh 1833: 11). Mr. G then further notes that there is an additional area of study, "an inquiry into the nature of those causes which have operated to produce its present state, and which may be designated as *Secondary Geology*" (Welsh 1833: 11). We see reflected here what became two important distinctions in geology (and which today are often represented in two different types of courses in the introductory undergraduate geology curriculum): *historical geology*, concerned with reconstructing the geological past history of the earth from its formation to the present day, and *physical geology*, a study of the processes that shape the earth, both in the past and the present (Laudan 1987).

The practitioners of geognosy, the careful observers of geological formations, realized that there were distinctive layers in rock formations that, although clearly different from each other, were reproducible in general properties from one geographic area to another, especially in the order in which these layers could be found. By 1830 four such divisions—termed *primary* or *primative*, *transition*, *secondary*, and *tertiary*—were commonly referred to in the geological literature. Two other types of rock formations were also identified, layers of rock deposits termed the *diluvium* and *alluvium*, the former in particular becoming controversial in that it hinted at potentially chaotic and catastrophic events. By the mid-1800s, many of the names associated with the modern geological time scale, including Silurian, Cambrian, and Carboniferous, came to be associated with distinct layers found in the rock record, as seen in Fig. 1.1 (Gregory 1921).

But the identification of these layers, and a chronological reconstruction of the earth's history, did not depend on the rocks alone. *Paleontology*, the study of the fossil

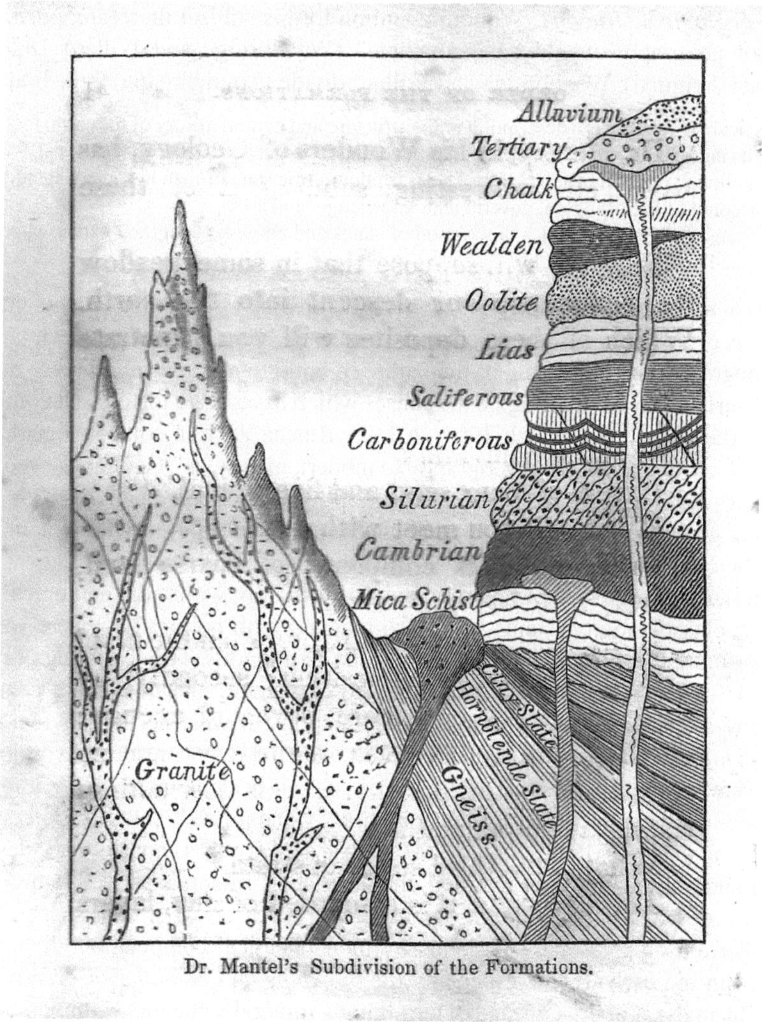

Fig. 1.1 Stratigraphic column. Reproduced from Delia W. Godding (1847) *First Lessons in Geology*

record, and a particular subdiscipline *Conchology*, the study of fossilized mollusk shells, also played a central role. It was therefore common for geology texts in the early 1800s to integrate these subjects within the larger discussion of the history of the earth.

It was also during this time that the relationship between geology and geography was being negotiated, especially as it involved the obvious overlap between the two fields, as they could both lay claim to a study of the surface of the planet. In her *First Lessons in Geology*, Delia Woodruff Godding defines geology as "a science which treats of the earth's structure, and of the substances which compose it" and differentiates it from geography, which "tells how the earth is divided into oceans and continents, mountains, plains, vallies [sic], &c., and what their names are; but Geology tells *how* and of *what* they are made" (Godding 1847: 9–10). But as noted in the influential work *Universal Geography* (1824), "The surface and soil of a country, the nature of its waters, and to some extent, its climate and salubrity, *depend on its geological structure.* Without attention to this subject therefore, our knowledge of geography would be incomplete" (Woodbridge and Willard 1824: 16; emphasis original). Woodbridge and Willard divide geography into three branches:

> Physical Geography is a description of the structure and natural history of the earth, including its natural divisions, climates, and productions.
>
> Political, or Moral Geography, is a description of the state of men in society, including an account of their religion, government, knowledge, and arts.
>
> Statistical Geography is a description of states and empires, with their extent, population, and resources. (1824: 1)

Physical geography, or *physiography*, is defined by Grace Anne Prestwich in her 1883 essay "The Face of the Globe" as "a science lying midway between geology and geography, and partaking of the nature of both, to account for all these diversities of surface, and to investigate the causes which have led to the present configuration of the globe" (Prestwich 1901: 201–2). Rosina Zornlin compares geology to ancient history and physical geography to modern history, or "Geology, or the Earth in its former condition; and Physical Geography, or the Earth as it is" (1839: 2). As physical geography clearly has a geological component to it, it will be included as a geological science in this volume.

But definitions were not the only areas of difficulty within the geological community; classification systems were also plagued with politics. Nowhere was this perhaps more clearly evident than in the study of minerals. The collecting of minerals and fossils was a popular hobby in the eighteenth and nineteenth centuries, whereby collectors would display their collections in what became known as the *natural history cabinet*. But such collections were not only considered works of art, they were works of science as well, and as such, it became increasingly important that the cabinets reflected some scientific, standardized classification system (Greene and Burke 1978; Simon 2002). Just as Carolus Linnaeus had classified plants and animals by their external characteristics, minerals as well had obvious visible properties, including color and, in the case of crystals, their shapes. There was also reason to classify minerals by their chemical composition. Technology utilized in the early 1800s, especially the *blowpipe* and *goniometer*, greatly aided both the analysis of crystals and determining a mineral's chemical composition.

At its most basic, a blowpipe is a thin tube into which a person blows, directing air over a flame, thereby increasing the temperature and intensity of the flame. A small amount of a mineral is subjected to this flame, and the color of the flame, as well as whether or not the mineral sample melts or fuses in the temporarily increased heat of the flame (or any other noticeable reaction) are properties used to identify the mineral (Newcomb 2009).

The simplest goniometers (instruments that measures the angles at which the planes of a mineral's crystals meet) are little more than glorified protractors, as shown in Fig. 1.2. In 1809 William Wollaston (1766–1828) developed an improved technology that was both more precise and accurate, in which reflections from the mineral's crystal planes are observed in order to measure the angles. This *reflective goniometer* was considered at the time to be "of the highest value to Crystallography" but there was "an opinion of its use being attended with some difficulty" (Brooke 1823: 27) In response, numerous books on the subject included detailed instructions in order to make this vital technology more widely used.

But while both crystals and chemical composition could be analyzed, they often gave conflicting suggestions as to how to classify minerals. Minerals with similar crystalline forms may have very different chemical compositions, while chemically identical minerals might manifest different crystalline forms. There resulted various schools of thought as to how to best deal with this complex situation. Crystallographer René Just Haüy (1743–1822), himself an amateur botanist, advocated a decidedly

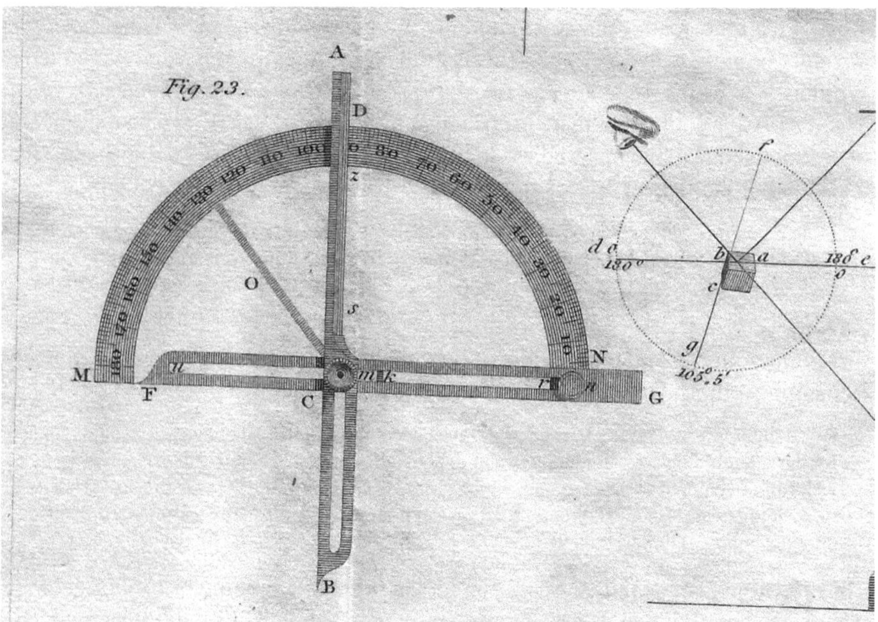

Fig. 1.2 Goniometer. Reproduced from Delvalle Lowry (1822) *Conversations on Mineralogy*, Amer. Ed

Linnaean system, where minerals were divided into four classes, the acidiferous, earthy, nonmetallic, and metallic substances. Despite the fact that his classification was heavily dependent on crystalline forms, these classes also incorporated chemical analysis. In contrast, Abraham Werner's system utilized external, directly observable characteristics (Greene and Burke 1978). Similarly, although Werner's system relied on observable external properties in identifying a mineral, he also admitted that chemical properties were important in mineral classification (Laudan 1987). It is therefore not surprising that some authors of geology textbooks and popularizations developed their own systems of classifying minerals, for example Delvalle Lowry in her *Conversations on Mineralogy.*

Another important area in which the geology of the nineteenth century differed from that of today is the modern distinction between amateur and professional (and the hierarchical value judgment these terms generally imply). Given the fact that geology was not widely taught in universities, most practicing geologists were self-taught (or were taught through attending public rather than formal university lectures). Similarly, paying positions at universities were uncommon; geology was rather an avocation than a vocation, and most geologists had to support themselves through writing, lecturing, fossil and mineral hunting, family fortunes, or other means. For example, Sir Charles Lyell himself only held a professorship in geology (at King's College) from 1831 to 1833. A modern delineation between professional (university-trained) and amateur (self-taught) began to develop gradually, and a true demarcation became established only around the time of World War I (O'Connor and Meadows 1976).

Finally, it must be noted that the term *scientist* itself was not widely in use through most of the nineteenth century. Commonly used terms instead included *natural philosopher* or *man of science*. The genesis of the term scientist is generally credited to William Whewell's 1834 review of Mary Somerville's *On the Connexion of the Sciences*. After noting that there is no "name by which we can designate the students of the knowledge of the material world collectively," Whewell (1794–1866) describes a meeting of the British Association for the Advancement of Science at which

> some ingenious gentleman proposed that, by analogy with artist, they might form scientist, and added that there could be no scruple in making free with this termination when we have such words as sciolist, economist, and atheist—but this was not generally palatable. (Whewell 1834: 59)

In his outgoing presidential address before the American Astronomical Society, Benjamin Apthorp Gould (1824–96) also lamented that

> By an unhappy, though perhaps natural, mischance, the English language has had no name for the scientific investigator, nor word to denote his calling….Therefore it was, that twenty years ago I ventured to propose one, which has been slowly finding its way to general adoption; and the word scientist, though scarcely euphonious, has gradually assumed its place in our vocabulary. (Gould 1869: 9)

Thus the nineteenth century was a pivotal yet volatile era in the history of geology, where even individual words themselves were disputed. But the disagreements between geologists concerning scientific vocabulary and the classification of minerals were just the tip of the iceberg in terms of scientific disputes in this rapidly evolving science. Complicating the geological debates of the time (including the nature of the forces behind the creation of large-scale geological features and the different types of rocks) was the fact that religion, not only empirical science, was often behind some of the proposals.

1.2 The Neptunism/Plutonism Debate and the Role of Religion in European Geology

The late eighteenth and early nineteenth centuries featured a series of passionately debated rival hypotheses as geologists attempted to interpret and explain the rock samples and geological structures they increasingly collected and studied. At the same time, the vast majority of these same scientists were practicing Christians, and the growing number of apparent disparities between the literal word of the Scriptures and the observed properties of the natural world provided a source of tension for both individual geologists, and for geologists and theologians in general. At the two extremes were those who felt that science owed nothing to religion in the search for truth, and those who believed that the literal and infallible Word of God always trumped the observations and hypotheses of the limited human mind. In the middle were the vast majority of geologists, who sought to discover the laws of the natural world (laws that they still believed ultimately derived from God), no matter how contrary these laws might appear to a literal reading of the Bible, or who strived to either find agreement between a more liberal interpretation of the Bible and the observations of geology or used geology in an attempt to prove the truth of events within the Bible. The situation was so volatile that an anonymous book reviewer lamented in 1832 that "Enveloped in visionary theories, the formation of the earth remained a mystery only to be speculated upon by the learned; whist the absurd notion, that all natural truth was contained in the volume of revealed religion, checked inquiry and forbade investigation" (*The Athenaeum* 1832: 77).

The first important scientific debate centered around whether water or fire was the main driving force behind geological activity. In the 1780s Abraham Werner interpreted the vast layers of rock called geological strata as being the result of an original primeval ocean that had covered the infant earth. In his hypothesis separate layers of different rock types formed in sequence as materials of different varieties were deposited one on top of the other. He therefore posited that over time deposition and chemical precipitation were responsible for all types of rocks. His model became known as the *Neptunist hypothesis* and was quickly embraced by a number of deeply religious geologists, in part because it appeared to parallel the Biblical account of the great Noachian Flood. However, basalt and similar rocks were found to be associated with volcanic features, which supported a rival hypothesis that posited that the earth's internal heat and the resulting melting of rocks drove geologic processes (the so-called *Vulcanist hypothesis*).

In his seminal 1788 treatise *Theory of the Earth*, Scottish geologist James Hutton (1726–1797) greatly extended the Vulcanist hypothesis to include the idea that, while some rocks were laid down by water, both heat and pressure from the interior of the earth uplifted and deformed the strata, creating many of the large-scale geological structures found in the environment (such as mountains). Hutton's model is therefore sometimes termed the *Plutonist hypothesis*, although in many nineteenth century texts the term Vulcanist is used synonymously (Hallam 1989). In addition, Hutton posited that the processes that formed geological structures in the past are precisely those that are still observed to modify the surface today. No supernatural events are required to either create mountains nor wear them away, only the slow, relentless working of physical processes. Not only was the Hand of God seemingly removed from the face of the earth in Hutton's model, but there was another serious problem for Biblical literalists: in order to explain the geological structures seen today, these slow processes required millions of years (or longer) to shape our planet, not the 6000 years of the Mosaic chronology. Also problematic in Hutton's *steady-state* or *uniformitarian model* was that there was no direction in the geological evolution of the earth. Endless cycles modified rocks and rock formations without apparent plan or design. "I can find no traces of a beginning, no prospect of an end," Hutton explained, in a seminal quote highlighted by Charles Lyell in his *Principles of Geology* (Lyell 1830: 63).

There was a backlash unleashed upon Hutton, along with charges of blasphemy and atheism. Richard Kirwan (1733–1812), chemist, mineralogist and President of the Royal Irish Academy, was one of Hutton's most vociferous public critics, and in the introduction to his *Geological Essays* (1799) argued that "*sound* geology *graduated* into religion, and was required to dispel certain systems of atheism or infidelity, of which they had had recent experience" (qtd. in Lyell 1830: 68; emphasis original). Geologists leaning toward a more literal interpretation of the Bible not only embraced Werner's Neptunist model for its emphasis on a directional progression of creation and the effects of worldwide floods, but bolstered it with their interpretation of the results of the rock and fossil studies of French anatomist and zoologist Baron Georges Cuvier (1769–1832).

Cuvier had spent many years studying fossil vertebrates, especially fish, and published a series of papers in the *Annales du Muséum*. Upon completing this work in 1810, he turned to synthesizing the results of his research and interpreting what he felt the evidence in the fossil record implied for the history of the earth. His 1812 work *Ossemens Fossiles* contained his earlier papers on fossil fish, and was also prefaced by an essay, *Discours Préliminaire* (Preliminary Discourse), that was later reprinted on its own as *Discours sur les Révolutions de la Surface du Globe* (Discourse on the Revolutions of the Surface of the Globe). Since delicate fossils were often found intact, Cuvier reasoned that they must have formed in situ relative to where the organism had died. But the fossil record demonstrated that there had been a series of widespread changes in the fossil types. Since Cuvier did not find evidence of intermediate or transitionary forms between the organisms in successive strata, he interpreted the fossil record as evidence that the planet had suffered a series of widespread (but, importantly, not necessarily worldwide) catastrophic flooding events. In the hypothesis known as *Catastrophism*, all life became extinct with the end of each geological catastrophe, and new forms of life were generated afterwards. Cuvier himself did not offer a model for the creation of these new species, but it is known that he was not a supporter of Lamarckian evolution (Hallam 1989; Rudwick 1997).

Nor did Cuvier invoke a religious explanation, despite the co-opting of his scientific ideas by a segment of the geological community in order to bolster a literalist interpretation of the Book of Genesis. The first problem was that Cuvier had not insisted on worldwide catastrophes (in conflict with the story of Noah); the second was that Cuvier had a reputation for being a religious skeptic. While he mentioned the Noachian Flood in his popular lectures, biographer Martin Rudwick argues that it was a tactic used to proactively defend his science against any accusations from Biblical literalists (Rudwick 1997).

The misleading association between Cuvier's work and Scriptural literalism is due in large part to the English translation of *Discours Préliminaire* by University of Edinburgh Professor of Natural History and founder of the Wernerian Society Robert Jameson (1774–1854). In what he renamed *Essay on the Theory of the Earth* (1813), Jameson made it clear in his preface and editorial notes that he believed that Cuvier had provided incontrovertible scientific evidence in support of a historical Noachian Flood. This interpretation was embraced by many of his readers, including some of the more religious British geologists (who, coincidentally, were also members of the clergy). This gave birth to the *Diluvianist movement*, a coterie of geologists who attempted to use the diluvial layer of the fossil and rock record to provide scientific evidence in support of not only the Biblical Deluge, but the entire Book of Genesis (Hallam 1989). The Catastrophist movement in general was also widely supported by religious literalists, as the genesis of new species after each catastrophe was seen to only be possible through a special creation by God.

One of the most widely respected, read, and cited diluvianists was William Buckland (1784–1856), originally Reader of Mineralogy at the University of Oxford. When he became Reader of Geology in 1819, his inaugural lecture (published as *Vindiciae Geologicae; or, the Connexion of Geology with Religion* [1820]) was "an attempt to shew [sic] that the study of geology has a tendency to confirm the evidences of natural religion; and that the facts developed by it are consistent with the accounts of the creation and deluge recorded in the Mosaic writings" (Buckland 1820: ii). As with other scientific works of his time, Buckland took the occasion to point out that the distribution of minerals and other niceties of the natural landscape were evidence of the benevolence and design of God, and argued against materialism by affirming that even though geological processes work via the natural laws God Himself set out, they still require the continual Hand of God to control and direct the occasional current-day catastrophe according to His design and plan (Buckland 1820). Buckland then sets out his central argument, including the following points:

1. The oldest rocks have no visible fossils; therefore life had a definite beginning, which was accomplished by the Hand of God.
2. The diluvial layers provide "so decisive and incontrovertible" evidence of a worldwide flood that "had we never heard of such an event from Scripture, or any other authority, Geology of itself must have called in the assistance of some such catastrophe to explain the phenomena of diluvian action … at a period not more ancient than that announced in the Book of Genesis" (Buckland 1820: 23–4).

But in order to successfully align geology with a Biblical interpretation, Buckland had to work out apparent contradictions as well as caution against common improper interpretations of the geological record. Most common of these was the interpretation of the fossil seashell-laden layers found on mountains as evidence of the Mosaic flood. The most obvious apparent contradiction between geology and theology, the short time frame contained in the Mosaic history, could be overcome by a variety of methods that involve reinterpretations of Scripture. These include the insertion of a gap of time between the beginning of the world and the six days of creation (i.e. between verses 1 and 2 of the Book of Genesis) or the interpretation of the six days of creation as far longer periods of time. It became common in the literature to refer to the six days as "six successive epochs of indefinite duration" (Bakewell 1828: 26), thus providing sufficient time for natural geological processes ranging from catastrophes to the gradual erosion of rivers to create all structures that were formed before the Great Flood. A third possibility is that there was a long period of time between God shaping the earth (separating the waters from the land) and the creation of life. Buckland offers a fourth hypothesis, that the lengths of the days of Creation is not 24 h. He explained that either of the last two would remove issues between geology and the Bible (Buckland 1820: 24). Not all of these explanations were original to him, but this summary proved useful to later authors, who would widely cite him.

While most mainstream geologists of his day lauded Buckland's reconciliation of geology and Scripture, some Biblical literalists were inclined to force geology to bow to the Divine Word. Their works became known as *scriptural geology* or *Mosaic geology,* named for George Bugg's *Scriptural Geology* (1826) and Granville Penn's *Comparative Estimate of the Mineral and Mosaic Geologies* (1822). Penn (1761–1844) was neither a geologist nor a member of the clergy; rather, he was an assistant chief clerk in the War Office and philologist who specialized in early Christian, Classical, and Byzantine literature (O'Connor 2007). He disparaged both the models of Werner and Hutton under the moniker of "mineral geology" and dubbed them "falsehood" to the truth of the literal Bible (Penn 1822: 5-6). His model of Mosaic geology followed directly from Scripture—the six days of creation were six 24-h days, the world was approximately 6000 years old, and all important geological structures derive from two acts of God: the original parting of the waters from the land (accomplished before life was created, thus explaining the lack of fossils in the lower strata), and the Noachian Flood. Penn based much of his argument on the philosophical notions of Isaac Newton, in particular the argument that not only is God the original Prime Cause, but that the laws of nature, as secondary causes, are insufficient to keep the universe running without the continued direct intervention of God.

The extinction of numerous species of animals reflected in the fossil record might be seen as problematic to Biblical literalists who consider the deity to be omnipotent and benevolent. Penn addressed this issue directly, arguing that God ordained which animals were to become extinct: "He who planned and regulated the *creation* of the earth, unquestionably planned and regulated also its *renovation*; and the *extinction of certain animal species*, which existed prior to that last revolution, is proved by their *exuviae,* to have been *a part of His plan in the renovation*" (Penn 1822, 342–3; emphasis original). Penn then argues that the special creation of modern species of animals by God (including those not mentioned in the Bible) is a logical assumption, as God already had to specially create new plants to cover the landmass after the receding of the flood waters.

Without the millions years needed to form rocks through geological processes, Penn's model instead suggested that just as God had made Adam and Eve fully formed—fully grown with their organs formed and bones hardened—the rock strata were similarly formed in their present state, rather than hardening from lava or being deposited one particle at a time. The biological and geological processes seen at work in the world today were not those by which God created the world and its creatures and rocks. Penn attempted to explain rocks that are constructed from older parts (such as conglomerates made of older pebbles and coal made of organic matter) as being the result of the Deluge of Moses. In one of the most curious aspects of his model, the current mountains and continents once formed the ocean floor, which rose after the Flood. The original rocks were somehow able to change places with the previous sea floor and formed a new basin in which to store the immense volume of water required to cover the entire earth in the Deluge.

Penn's work was panned by a number of influential geologists and theologians, but enjoyed some success with the general public. Adam Sedgwick (1785–1873), a clergyman and Professor of Geology at Cambridge, wrote in his *Discourse on the Studies of the University* (1834) that Penn and his colleagues demonstrated a "shameful want of knowledge of the fundamental facts… dishonored the literature of this country… overlooked the aim and end of revelation, tortured the book of life out of its proper meaning, and wantonly contrived to bring about a collision between natural phenomena and the word of God" (qtd. in Hitchcock 1835: 266).

Catastrophism and Neptunism, both of which were deemed relatively friendly to a reconciliation between geology and Scripture, remained popular in Europe through the first few decades of the nineteenth century, but eventually gave way to Uniformitarianism and Plutonism, after the publication of *Principles of Geology*, the seminal work of geologist Charles Lyell (1797–1895). Based on a decade of rock and fossil field work across Europe, Lyell's three-volume work laid out the evidence for a Huttonian model of continuous, gradual geological processes continuing to shape the earth's surface over long periods of time. Lyell explained geological structures though observed processes active in the world today, without the intervention of any supernatural agency. He, like Hutton, did not see evidence of a beginning in the geological world, and was careful to distinguish between geology and *cosmogony*, the scientific speculation about the origins of things. Like Hutton, Lyell was attacked for removing God from this central role, and critics questioned his religious faith. In response, in both the last chapter of volume 3 of *Principles of Geology* (1833) and his 1832–33 public lectures on geology, he affirmed his own belief in a powerful God who created man and all living things according to a great plan, and added that "To assume that the evidence of the beginning or end of so vast a scheme lies within the reach of our philosophical inquiries, or even of our speculations, appears to us inconsistent with a just estimate of the relations which subsist between the finite powers of man and the attributes of an Infinite and Eternal Being" (Lyell 1833: 385). Lyell's pious statements did little to soften the blow his geological studies struck against literalist interpretations of the Bible. Those who subscribed to the arguments of Buckland were less threatened, but many of them were still reluctant to abandon the power of catastrophes and floods and embrace the slow and steady processes of the modern geological world, despite the preponderance of evidence to support Lyell's model. These gentlemen formed the authoritative body of European geology, and were both heavily influential upon, and were heavily cited by, all popularizers of geology, both in England and in America, and especially women writers, who attempted to convince their critics that they had the requisite background to speak about science by citing and quoting these illustrious men of science.

1.3 The Religion of Geology: Silliman, Hitchcock and New England Geology

Attempts to reconcile geology with Scripture were especially popular in America, partly due to the deeply held religious beliefs of many American geologists. This enhanced and prolonged the popularity of Catastrophism and Neptunism in the U.S. throughout the first half of the nineteenth century (Matijasic 1987: 416). The result was a specific type of reconciliation between geology and theology that took the work of Buckland an important step further. Not only was there no contradiction between the two, but geology was seen to provide proof of the events in the Bible, and thus the study of geology, even more so than the contemplation of astronomy, would deepen faith and devotion. This movement became known as the *religion of geology*, a term coined in an 1835 article by one of its main proponents, the Reverend Edward Hitchcock, Professor of Geology at Amherst College. Hitchcock and his friend and former geology instructor, Yale College Professor Benjamin Silliman, greatly influenced American geology during the mid-1800s through their widely-read articles, books, and popular public lectures.

Benjamin Silliman (1779–1864) was a native of Connecticut and grew up in a devout Congregationalist family. He received bachelors and masters degrees from Yale College, and passed the bar in 1802. At the suggestion of Yale President Timothy Dwight he studied chemistry in order to become the university's chemistry professor, and in addition studied geology in Edinburgh in 1805–6, a time of intense debate between adherents of the geological ideas of Werner and Hutton. While he was exposed to all points of view, he favored those of Werner to varying degrees throughout his life, although he eventually espoused a view that modified and synthesized all of these disparate models. For example, in his 1829 *Outline of the Course of Geological Lectures Given in Yale College*, he explained that although it was "the fashion of the day to attribute almost every thing in the earth to igneous agency, I shall probably be thought behind the present state of opinion, while I maintain, that the chemical affinities, through the medium of aqueous solutions of the great chemical agents—as well as of water itself, have also produced important effects in the early arrangements of the planet." His argument was that even though there was irrefutable evidence that igneous processes were important in shaping the earth, scientists should not "exclude any of the great powers, which we find in actual operation; or, of whose ancient activity there appears probable evidence" (Silliman 1829: 3–4). By affirming the power of water, and more especially the actions of water in the past, Silliman kept the door wide open for a scientific discussion of the Mosaic Deluge.

Silliman studied and taught geology at Yale, using the rich geological structures of New England as his laboratory. He widely disseminated scientific knowledge through his own writings and university and public lectures, as well as through his work as the creator and editor of the *American Journal of Science and Arts* (often called *Silliman's Journal*). But religion was ever-central to his geological studies, and was interwoven into his scientific writings and presentations. For example, he opened his introductory lecture with the explanation that "Among the privileges which the Deity has conferred on man, one of the greatest is the capacity and the disposition to acquire knowledge" (Silliman 1828: 5). He then listed various types of human scientific knowledge, beginning with "moral science," which he counted as the "most important branch... which informs us of our condition, as moral and accountable beings, and of our relation to the Creator" (Silliman 1828: 6). Near the end of his lecture, he finally arrived at geology, which he stated "abundantly confirms the truth of the Mosaic account of the creation, and of the succession of its various epochs, and it exhibits, in the arrangement and contrivance of the mineral strata, decisive proofs of the power, wisdom and design of its author" (Silliman 1828: 32).

Silliman's geology classes also included numerous references to the Mosaic tradition and how geology could be used to confirm its truth. As in the case of Buckland and the English reconciliation tradition, Silliman openly addressed apparent contradictions between geology and the Biblical chronology and gave his opinions on the various solutions provided by both Biblical and scientific scholars. In the preface of his *Outline of the Course of Geological Lectures Given at Yale College* (1829: 7) Silliman asked the question "Are the discoveries of geology consistent with the history contained in the book of Genesis?" His answer was unequivocal: "Respecting the deluge, there can be but one opinion... geology fully confirms the scripture history of that event" (Silliman 1829). He explained that although the Bible is "not a book of physical science... there are two great events recorded in it, which, although they have a momentous moral bearing, are, *in their nature*, entirely physical; we allude to the creation and arrangement of the planet, and to the deluge which was made to sweep over its surface" (Silliman 1829: 7).

In the same work Silliman (1829: 4) claimed that he was "neither Wernerian nor Huttonian, Neptunian nor Plutonist." He invoked water and fire as needed, and drew upon both catastrophic and slow, gradual processes as needed to explain the particular geological subject at hand. For example, he explained that the "creation of the planet was no doubt instantaneous, as regards the materials, but the arrangement, at least of the crust, appears to have been gradual," but then added a further religious twist by stating that the order of the important steps in this gradual process "corresponds with the *order of the events* narrated in sacred history" (Silliman 1829: 67). Silliman further separated himself from the scientific lineage of Hutton (and later Lyell) in the important respect of openly discussing the creation of the earth, in keeping with his Wernerian leanings (and the perceived natural sympathy between Neptunism and Scripture).

Throughout the body of his *Outline*, Silliman used geological examples to illustrate the benevolent and well-designed plan of the Deity in preparing the planet for its eventual habitation by humans, and affirmed that although the natural world depends on natural laws (which were created by God), God is ever-present in the world to this day. As Buckland before him, Silliman warned against incorrectly attributing the older geological strata to the Noachian flood, and devoted an entire 30-page section to outlining the physical properties of the Deluge, including the depth of the water, its effects (including the extensive diluvia which he saw with his own eyes in the Connecticut Valley), and the source of the extraordinary volume of water required to cover the entire earth. Silliman also used the universality of flood myths around the world as further evidence of the veracity of the Deluge. In his discussion of the geological evidence for the Great Flood, Silliman specifically addressed mastodons, whose fossils were increasingly found in the eastern United State in the early 1800s. He noted that some of the animals present on the ark may have afterwards become extinct through "unknown causes" and that mastodons might be among them. His personal opinion was that although "most of the mastodons perished at the deluge, I have no objection to admitting that some of them, whose skeletons are found, may have perished before, or since that event" and further explained that some may have "foundered in salt licks and marshes… miring, as cattle do at the present day" (Silliman 1829: 93).

Silliman also inserted his views on geology and Scripture into his editions of others' writings. For example, in 1839 Silliman edited the American edition of Robert Bakewell's *Introduction of Geology*. In an Appendix he laid out five solutions to the apparent discrepancies between geology and Scripture, especially the Mosaic chronology. These solutions include the use of periods of indefinite length to explain the six days of creation and a gap of time between the beginning and the first day. However, despite his deeply held religious beliefs, and his desire to use geology to provide evidence for them, Silliman remained rooted in geological theory, and in this Appendix explained that

> our convictions are confirmed by surveying, with Mr. Lyell… that since the creation, as regards geological causes, except their varying if not diminished intensity of action, all things remain as they were; no new code of physical laws has been enacted; while the beginning was with God, the continuation of events is with us, and a distant posterity may not witness their termination. (Silliman 1839: 546)

Silliman was also critical of Penn's Mosaic geology, although much more gentle in his criticism than Sedgwick and others. In his *Outline of the Course of Geological Lectures*, Silliman (1829: 77) noted that although his fellow Christian had "served the cause of truth," his work "has, in our opinion, left the question between the critics and the geologists embarrassed with all its difficulties." In a 1834 letter from English paleontologist Gideon Algernon Mantell (1790–1852) to his friend, Charles Lyell, Mantell marveled at the size of the enthusiastic audiences at Silliman's public lectures, and expressed wonder at Silliman's ability to uphold the geology of Lyell while simultaneously reassuring his god-fearing audience that the new geology did not contradict their Holy Book (Wennerbom 1999). Ironically, Silliman later edited the American edition of Mantell's *The Wonders of Geology* (1839) and included an introduction in which he noted that the geological record agrees with Scripture in the late appearance of humans in the historical record, and in the evidence of a deluge. However, by this time, Silliman had modified his opinion to include the possibility that the geological record might be as equally well described by numerous local deluges as one worldwide event.

This evolution in the interpretation of the geological (and scriptural) records would ultimately lead Silliman's most famous student to eventually (and with great reluctance) abandon the idea that geology could unequivocally prove the existence of the Noachian Deluge. Edward Hitchcock (1793–1864) was born in Deerfield, Massachusetts, and after suffering complications from the mumps at age 21 had a religious epiphany and fervently embraced his father's Congregationalist faith (Lawrence 1972). Two years later he became the principal of the Deerfield Academy, and was introduced to geology through a series of public lectures given by Amos Eaton. Between 1819 and 1821 he studied both theology and geology at Yale and worked in Silliman's laboratory. After spending four years as a minister at a church in Conway, Massachusetts (while simultaneously conducting geological field work), he was appointed the first Professor of Natural History at Amherst College in 1825, and then in 1845 became the president of that institution as well as Professor of Natural Theology and Geology. Hitchcock's reputation as a geologist was further bolstered by his appointment as State Geologist of Massachusetts in 1830 and the resulting surveys and maps he published of the geology of Massachusetts and other Northeast states. Among his most important and lasting geological contributions was his nearly thirty years of work on the fossils of the Connecticut Valley, especially dinosaur footprints and shale impressions (Little 2003).

Hitchcock's interest in the intersection between religion and geology dated back to at least 1823, to a sermon he delivered to mark the foundation of the Pittsfield Lyceum of Natural History. Entitled *The Utility of Natural History*, the lecture defined the role of the Christian geologist. According to Hitchcock, there were three utilities of natural history: in relation to the common interests of man, the intellectual improvement of man, and religion. The first includes the development of medicines and provision of natural resources, while the second leads to an appreciation of the perfection of nature and an improvement in the reasoning powers. The third, and clearly most important, is its relation to religion, in increasing feelings of devotion to God, making apparent God's design, demonstrating the existence and attributes of God, and elucidating (and proving evidence for) specific passages in the Bible, especially the Noachian flood. Hitchcock (1823: 21) explained to his audience that "in addition to that mass of evidence that flows in upon us, to prove the occurrence of such a catastrophe in former days, from the histories and traditions of almost every nation and tribe civilized and savage under heaven, we find the records of geology everywhere speaking the same language." After following Silliman and Buckland in warning against using the ancient strata as evidence of the Deluge, he laid out the evidence found in the diluvium, including that of the nearby Connecticut Valley, proclaiming that (1823: 23) "wherever the infidel goes, he finds the very stones crying out against him." After summarizing several common methods of reconciling the geological and Mosaic chronologies, he declared (1823: 26) that "there exists very little diversity of opinion—geologists of every name, infidel and Christian, generally agreeing, that the voice of nature unites with the history of all nations in testifying to the truth of the Scriptures that man has existed on the globe for not more than five or six thousand years." Therefore geological speculation about the history of the earth prior to the existence of man becomes "of little practical importance" (Hitchcock 1823: 26).

However, as Hitchcock continued to explore the geology of his home state and those around it, he found the rocks singing a slightly modified tune. While he never wavered in his belief that geological evidence supported his Christian beliefs, he found that he had to modify his arguments of exactly how that evidence fit into scriptural tradition. The first signs of this appeared in his 1833 report on the geological survey of Massachusetts. After studying the diluvium in both Massachusetts and New York in great detail, he found what he believed to be irrefutable evidence of more than one event, moving materials at different locations in slightly different directions. He explained to his reader that despite the fact that it might appear that this evidence could be contrary to the classic Mosaic tradition, it was only necessary that the latest event correspond to the Noachian event. An unknown (and perhaps unknowable) number of other events could have occurred in the long expanse of time between the origin of the planet and the time of Noah. With this understanding, "all apparent collision between geology and revelation vanishes" (Hitchcock 1833: 166).

In 1835, Hitchcock published a series of three articles in the *Biblical Repository and Quarterly Observer* in which he further explained his opinions on the power of geology to prove the existence, design, and benevolence of God. He began his argument with the explanation that "The principles of geology have long been regarded not only as hostile to revealed truth but as favourable to atheism." His intention was therefore to demonstrate that the contrary was in fact true, and "that the student of natural theology will find the records of geology no unfruitful source of evidence as to the existence, perfections and plans of Jehovah" (Hitchcock 1835: 113–4). After laying out his evidence (including proof that God still intervenes in the world by mitigating the effects of catastrophes), he ended with the following statement:

> Such is the religion of geology. Prejudice may call it atheism, because it presents before us views so new and peculiar; and scepticism may pervert these views to suit an unsubdued and unholy heart. But we call this religion a transcript of the Divine Perfections. And if there be one spot in the whole circle of science, where the student of natural theology can find the fuel to kindle up the flame of devotion, it is, as it seems to us, when he secures a live coal from the altar of geology. (Hitchcock 1835: 138)

Having defined his new interpretation of the relationship between geology and religion, he continued in the second article to explain in detail over a dozen different means of reconciling the Mosaic and geological timescales. This argument ended with the opinion that even if none of these methods were wholly satisfactory, given both the state of geological knowledge and the exegesis of the Book of Genesis, it would be "premature" to decide there were any "real discrepancy between them" (Hitchcock 1835: 327). The final article set out a series of coincidences between geology and revelation, including the beginning of the world, the importance of both water and fire in shaping the world, the existence of an early period in which the world was submerged in water (before the land was separated from the waters), the late appearance of humans, the Noachian flood, and the final end of the world in fire (Hitchcock 1835: 443–51).

Hitchcock followed this series of articles with another triumvirate in 1837–8 in which he set out evidence for the Great Flood (Hitchcock 1837; 1838) However, these articles demonstrate a further evolution in his thinking, from a worldwide flood reflected in the traditions of myriad cultures, to accepting Cuvier's interpretation of the diluvium as evidence for large but perhaps localized floods across the globe. In the final paper he came to the modified conclusion that "the Mosaic account of the deluge stands forth fairly and fully vindicated from all collision with the facts of science" yet openly resisted stating that "geology strikingly confirms the Mosaic history, as it has been customary to do in most of our popular treatises on the deluge" (Hitchcock 1838: 27). These basic concepts of his *religion of geology* were included in his 1840 textbook *Elementary Geology*, which included discussions of "Illustrations of Natural Religion from Geology" (revealing the existence, designs, and benevolence of God) and the "Connection of Geology with Revealed Religion" (1841: 275; 281). Much of his argument was a summary of his earlier articles on the subject. Importantly, Hitchcock issued a second edition the following year in order to include a discussion of the new theories of glacial action coming out of Europe, whose application to the problem of the diluvium he recognized would soon be regarded as a "new and important chapter in Geology" (Hitchcock 1841: vii). He further explained (1841: 218) that he felt the theory would be better named the "Glacio-aqueous theory" in recognition of the "vast deluges" which would be the result of the sudden melting of glaciers. Hitchcock was eventually proven correct in this belief a century later, when geologists discovered that the string of ancient lake segments that Hitchcock had found evidence for in the Connecticut Valley were created near the end of the last Ice Age (circa 15,000–18,000 years ago). In 1939 Dartmouth geologist Richard Lougee named the formation Lake Hitchcock in the Reverend's honor (Little 2003: 94).

Hitchcock continued to write about the connections he saw between geology and religion for the remainder of his life, with his most developed work, *The Religion of Geology* (1851), laying out the final evolution of his beliefs. There was no longer worldwide evidence for a Noachian flood, but the Deity was still required to create the world and all life and regulate the laws of nature. As Lawrence (1972) argues, while geologists had to cede the Noachian deluge to the onslaught of scientific data, with some new interpretations these same scientific facts could still be aligned with Biblical Scripture.

Therefore within the writings of this time period a variety of different negotiations between theology and geology can be seen. In the minority are Young Earth literalists, who subscribe to an earth of about 6000 years in age and a literal interpretation of the Biblical six days of Creation. Among the Old Earth geologists (who accept that the geological record demands an age in the millions of years, if not older), there are Diluvianists who hold that the geological record contains incontrovertible evidence of the Deluge (as well as the original separation of the waters in Genesis). As the century unfolds, an increasing number of geologists will say that either the evidence is consistent with the Deluge, rather than definitive, or will hold that the Deluge simply cannot be found in the record solely because it was of so short of a duration. There are those who merely note that God created the world and the laws of nature and do not discuss any particular Biblical events within their scientific writings. In addition, as explained there are myriad ways of explaining the six days of Creation in a metaphorical rather than literal sense (such as a gap of innumerable years between verses 1 and 2 of Genesis, or six innumerable ages).

This tradition of combining geology and religion (seen in both Europe and New England) played a central role in shaping the works of many female popularizers of geology, who widely drew upon and cited these male experts and authors. The ability of geology to further the cause of religiosity and morality played a significant role in opening up this space for women writers, where they were able to craft scientifically simplified and purposefully entertaining works for both children and other women without challenging the accepted societal roles for women. This opportunity was especially important, given the significant challenges faced by women who desired to become involved in the geological sciences in this era.

1.4 Challenges and Opportunities for Women in Geology

It is no accident that the term *men of science* was commonly used in the nineteenth century to describe what we would today call scientists, for science was very much considered the domain of men. Yet women did contribute to the scientific endeavor during this century, in meaningful, if not less visible, ways. Women were able to carve out well-defined niches for themselves, especially in the biological rather than physical sciences. In fact so many women took part in botany that it became considered a feminized science, resulting in a backlash later in the century when male practitioners reclaimed the science by professionalizing it and marginalizing the women who had been successful collectors, writers, and illustrators in the field just a few decades before.[1]

But it was in large part precisely the male-defined societal restrictions placed on women that had led them to be so successful in botany in the first place. Beginning in the late nineteenth century, the concept of separate spheres for men and women became highly ingrained within polite middle-class society, a philosophy that relegated women to the domestic sphere of the home. Middle-class homes often featured carefully tended gardens, in which women were encouraged to walk and enjoy with their children, children that not only played under their watchful eye, but asked questions about the world they observed within their gardens (and beyond). This led these same women to learn about botany specifically, and the natural world in general, and to share this knowledge with other mothers who likewise had inquisitive

[1] Ann Shteir (1996) describes the efforts of the professional botanical scientific community to take back the science from the hands of women as, in many ways, paralleling the shift in Victorian literary circles away from women novelists to a masculinization of that literary field.

children to help educate. As will be explored in Chap. 2, this led to a revolution in the publishing industry, and a rather lucrative venue for women with both scientific knowledge and literary talent.

But the domestic expectations for women also included providing companionship to their husbands, or in the case of unmarried women, to their widowed fathers or bachelor brothers. For this reason, it was socially acceptable for women to act as scientific assistants to their male relatives or husbands, even going so far as to accompany them on geologic expeditions (so long as the terrain did not prove a threat to the delicate feminine constitution). These female assistants played well-defined roles, often doing the less-glamorous and more time-intensive tasks such as cataloging samples and writing and illustrating field notes (Burek and Higgs 2007). As drawing was considered a proper skill for a lady to possess, girls were encouraged to take it up from an early age. As a result, it was not uncommon for the sketches and plates of fossils, landscapes, geological formations, and crystals in geological papers and books to be the handiwork of women. Additionally, in the 1800s women also made significant contributions to the state geological surveys in the United States as illustrators, including Orra White Hitchcock, wife of Edward Hitchcock (Aldrich 1982).

Fieldwork presented unique challenges to women; besides the societal limitations that their work generally be done under the watchful eye of a protective male relation, and only in areas that posed no undue physical stress to the lady, there was the very practical consideration of what a woman could wear in the field and how clothes would hamper a woman's ability to move (as well as detract from her general comfort). Fortunately, with the introduction of the divided skirt or bloomer in bicycling in 1851, a reasonable compromise between practicality and decorum could be achieved (Burek and Kölbl-Ebert 2007).

In reality, any outdoor activity could turn into a geological expedition. As women would take leisurely strolls across their family's lands, or along local coastlines, their eye would catch attractive mineral samples and, in particular locations, fossils, that could be collected, and often sold. This afforded women with not only a means of providing for their own and their family's financial needs, but participating in the process of scientific discovery itself. This was particularly true in the British Isles in the early nineteenth century (Kölbl-Ebert 2002). The earliest well documented example is Etheldred Benett (1776–1845), an unmarried woman who translated the security of her family's homestead and financial resources into a notable collection of local fossils (Laming and Laming 2007). More widely known are four unmarried female fossil collectors from Lyme Regis in Dorset, England, the Philpot sisters Mary (1777–1838), Margaret (d. 1845), and especially Elizabeth (1780–1857), and their friend Mary Anning (1799–1847). The Philpot sisters amassed an impressive collection of fossil fishes that came to the attention of famed geologist William Buckland, while Anning learned fossil hunting from her carpenter father, and became so successful that she was able to support herself and her widowed mother. Among her most important discoveries were the first icthyosaurus found in England (1811) and the complete skeleton of a new species, the plesiosaur (1824) (Creese and Creese 1994).

In Scotland, Lady Eliza Maria Gordon Cumming (c. 1798–1842) became well-known for her sizable collection of fossil fish, sharing her collection with such geological illuminaries as William Buckland, Roderick Murchison, and Louis Agassiz (Creese 2007), while Barbara, Marchioness of Hastings (1810–58) is known for the Hastings Collection of fossils, some purchased, some she collected herself, which now resides in the British Museum. While collecting fossils she made careful observations of the local strata, and published several scientific papers on her observations that included meticulously drawn section maps (Creese and Creese 1994).

A number of women also contributed to geology through their firsthand written accounts of geological events or structures. Travel writer and children's author Maria Graham (later Callcott) (1785–1842) included a first person account of the Chilean earthquake of 1822, in her *Journal of a Residence in Chile, during the year 1822* (1824), a description worthy of quotation by Charles Lyell in his *Principles of Geology* (Creese and Creese 1994). Likewise, American writer Hester Lynch Piozzi's fictionalized but scientifically accurate account of an Italian earthquake was reprinted by several magazines between 1789 and 1793. Finally, Mary Austin Holley integrated her observations of Texas geology into her *Texas. Observations: Historical, Geographical, and Descriptive* (1833 and 1836) while Susan Fenimore Cooper, daughter of the famed novelist, integrated detailed observations of the New York environment, including its geology, into her *Rural Hours* (1850) (Aldrich 1990).

Yet with all these avenues through which women could participate in science in general, and geology in particular, there were certain very influential doors that were firmly shut against them, namely the various learned societies. Excluding women from the official roles and meetings of scientific societies in Britain was the norm in the nineteenth century. A notable case is the British Association for the Advancement of Science (BAAS). Founded at a meeting in York in 1831, the 353 charter members were all male (although women were present at the meeting). Despite the bylaws' use of gender-neutral terms such as *persons* and *members*, the implicit intent of the meaning was men-only. For example, upon hearing that some of his colleagues planned to bring their wives to the 1832 meeting in Oxford, President-elect William Buckland voiced his opinion that their presence was appropriate only at private parties (Morrell and Thackray 1981). By 1833, special women's tickets were issued that allowed women access to the general meetings but not the sectional meetings where the majority of the scientific papers were read. The tickets became so popular that it was suggested that they be limited to members who had renewed their memberships. After further discussions, women were admitted to the 1838 meeting in Newcastle-on-Tynes, and nearly matched the men's attendance (1100 vs 1300). However, they were prohibited from attending the reading of the papers in the Botany and Zoological section due to some of the potentially inappropriate content. Not surprisingly, women were barred from the next few meetings, and their numbers were controlled for the 1841 meeting at Plymouth (Lonsdale 1970). The first woman member is thought to have been Miss Bowlby from Cheltenham, who was admitted in 1853 (Morrell and Thackray 1981).

Another example is the Royal Astronomical Society (RAS). In 1886, Elizabeth Isis Pogson (1852–1945), who had served as an unpaid astronomical and meteorological assistant for her father, Norman Robert Pogson, Government Astronomer at the Madras Observatory in India since 1873, was properly nominated for Fellowship of the RAS. Before bringing the nomination before the full membership, the Council asked for a ruling by its legal counsel as to whether their charter could admit women. When the opinion was given that a properly nominated woman had the right to be brought forward, the Council then sought a second opinion, which reported that because female Fellows had not been considered at the time of the founding of the Society, and the Charter only used male pronouns, it must be assumed that women were not meant to be included. In a politically masterful move that removed responsibility from their shoulders, the Council agreed to forward Miss Pogson's nomination, with the caveat that if she failed to achieve the necessary number of positive votes that it would not be a slight against her personally, only a signal that the issue of women in general should first be discussed in the General Meeting. As expected, the nominators withdrew her name to save her the embarrassment (Dreyer and Turner 1923). The first women were not elected to the RAS until 1916, and Elizabeth Isis Pogson (then Kent) was finally elected to the RAS in April 1920 (*Monthly Notices* 1920).

Fortunately, the development of amateur societies, such as the Liverpool Astronomical Society (LAS), founded in 1881, at least partially filled this void. Pictured as a "halfway resting-place between the amateur public and the Royal Astronomical Society" (Davies 1883: 269), at least eight of its 300 members in 1887 were women, including famed astronomy popularist Agnes Clerke (1842–1907). Highly-respected amateur observer Elizabeth Brown (d. 1899), a well-to-due spinster who was one of the only women in Britain to have her own observatory, served as Director of the Solar Section from 1883 to 1890 (Chapman 1998).

It is important to note that discrimination was not universal in the scientific sphere. The Zoological Society of London and Royal Entomological Society admitted women from their inceptions (1829 and 1833), and the Botanical Society of London and Horticultural Society admitted women in the 1830s (Morrell and Thackray 1981). The Royal Geographical Society voted to admit women in 1892, but had accepted papers from women decades earlier. Women were prevented from joining the Geological Society of London until 1919 but, as in the case of astronomers, were able to join local societies and the Geological Association (Kölbl-Ebert 2002). The Linnaean Society of London and Royal Microscopical Society finally opened up full participation to women in 1905 and 1909, with the Physiological Society and Chemical Society admitting female members in 1915 and 1920, respectively.

But nowhere was this discrimination more blatant and persistent than in the prestigious Royal Society of London, founded in 1660. Women were omitted from its original charter and statutes as well as its meetings as a matter of practice, with few exceptions. In the mid-nineteenth century, the Royal Society introduced *conversaziones*, events where guests invited by the President would be entertained by scientific displays appropriate to a social audience. In 1876 these events were opened up to women, and at an 1899 event Hertha Marks Ayrton (1854–1923), a well-known expert on the electric arc, demonstrated her research (Mason 1995). A graduate of Girton College, Hertha married physicist William Ayrton in 1885 after attending his classes at the Finsbury Technical College and became interested in his research. Prior to her 1899 presentation for the RS she had published 12 articles in the *Electrician*, had presented an invited paper to the Institution of Electrical Engineers (the first paper presented by a woman) and became that society's first woman member. In 1899 she also presided over the Physical Science section of the International Congress of Women in London and soon after presented a paper at the International Electrical Congress in Paris. In 1901 her paper "The Mechanism of the Electric Arc" was read before the Royal Society by Fellow John Perry and her ground-breaking book, *The Electric Arc*, appeared the following year (Mason 1995). In 1902 she was duly nominated for Fellowship in the Royal Society, but the Council sought advice from their legal representation because of the use of the male pronouns in their statutes. Their attorneys ruled that, in their opinion, married women were ineligible for Fellowship and it was doubtful that unmarried women were eligible under the current statues (Mason 1991). The ineligibility of married women conveniently negated Ayrton's nomination without further discussion, and was in keeping with British common law, which held that a woman's person was covered by her husband or father, therefore making membership on her own right impossible.

The 1919 Sex (Disqualification) Removal Act prohibited private societies and universities from using antiquated language in their charters to bar women, but not all societies rushed to embrace the fairer sex. Caroline Haslett of the Women's Engineering Society wrote a letter to the Royal Society Council in 1922 specifically inquiring if women were eligible for membership under the 1919 Act. Nearly three years later she received her reply, that women were indeed eligible under the present charter if they possessed the proper scientific credentials. But as no women candidates were nominated, the issue was never tested, and the first women Fellows were only elected in 1945 (Lonsdale 1970).

But while official membership in learned societies was out of reach for women in many scientific fields, scientific knowledge could be obtained through attending public lectures offered by some of the most well-known scientists of the time. Some of these were permanently housed at particular locations (such as the Royal Institution) while others were traveling series of lectures by such intellectual giants as Humphrey Davy and Michael Faraday in England, and Yale Professor of Chemistry and Mineralogy Benjamin Silliman in the United States. Public lectures aimed at a non-specialist audience were particularly fostered by the so-called lyceum movement beginning in the 1820s, not only in the UK where it originated, but the United States. While he was not the founder of the American lyceum movement, one of its most ardent supporters, especially in New England, was Josiah Holbrook (1788–1854), who had served for two years as Silliman's lab assistant at Yale.[2]

Supported by both benefactors and fees, lyceums provided lectures and demonstrations in order to increase the scientific and technical proficiency of workers in order to serve the economy. But since the lectures were open to all, teachers and older school children could benefit as well. By 1830 there were lyceums in a hundred New England towns and cities, the largest concentration being in Massachusetts (Bode 1956). A decade later, attending public science lectures had become so

[2] For more on the Lyceum movement in America, see Bode (1956).

popular in New England that it was not unheard of for two thousand people to attend one of Benjamin Silliman's lectures at the Lowell Institute in Boston, many of them women (Rossiter 1971).

While the lectures at the Royal Institution, lyceums, and other venues provided opportunities for women to increase their scientific understanding, it became clear that there was a market for published background material that would explain the lectures and experiments to the attendees as well as allow them to revisit the concepts at a later date. Hence, British (and later American) authors began producing popularized science books aimed at a non-technical audience, especially children and women. Written in an easily accessible language (in sharp contrast to the treatises found in professional science journals), their aim was to both amuse and inform. The need for popularizations of science opened an important door for women who stood at the periphery of science, and allowed them to contribute in a very meaningful (although limited) way.

References

Aldrich ML (1982) Women in paleontology in the United States 1840–1960. Earth Sci Hist 1(1):14–22

Aldrich ML (1990) Women in geology. In: Kass-Simon G, Farnes P (eds) Women of science: righting the record. Indiana UP, Bloomington, pp 42–71

Bakewell R (1828) An introduction to geology, 3rd edn. Longman et al, London

Bode C (1956) The American lyceum: town meeting of the mind. Oxford UP, New York

Brooke HJ (1823) A familiar introduction to crystallography. Phillips, London

Buckland W (1820) Vindiciæ geologicae. Oxford UP, Oxford

Burek CV, Higgs B (2007) Introduction. In: Burek CV, Higgs B (eds) The role of women in the history of geology. The Geological Society, London, pp 1–8

Burek CV, Kölbl-Ebert M (2007) The historical problem of travel for women undertaking geological fieldwork. In: Burek CV, Higgs B (eds) The role of women in the history of geology. The Geological Society, London, pp 115–122

Chapman A (1998) The Victorian amateur astronomer. Praxis, Chichester

Creese MRS (2007) Fossil hunters, a cave explorer and a rock analyst: notes on some early women contributors to geology. In: Burek CV, Higgs B (eds) The role of women in the history of geology. The Geological Society, London, pp 39–49

Creese MRS, Creese TM (1994) British women who contributed to research in the geological sciences in the ninteenth century. Br J Hist Sci 27(1):23–54

Davies WH (1883) To the editor of The Observatory. Observatory 6:269

Dreyer JLE, Turner HH (1923) History of the Royal Astronomical Society 1820–1920. RAS, London

Godding DW (1847) First lessons in geology. Parsons, Hartford

Gould BA (1869) Address. Proc Am Assoc Adv Sci 18:1–37

Greene JC, Burke JG (1978) The science of minerals in the age of Jefferson. Trans Am Phil Soc 68(4):1–113

Gregory HE (1921) History of geology. Sci Mon 12(2):97–126

Guntau M (2009) The rise of geology as a science in Germany Around 1800. In: Lewis CLE, Knell SJ (eds) The making of the Geological Society of London. The Geological Society, London, pp 163–177

Hallam A (1989) Great geological controversies, 2nd edn. Oxford UP, Oxford

Hitchcock E (1823) Utility of natural history. Allen, Pittsfield
Hitchcock E (1833) Report on the geology, mineralogy, botany, and zoology of Massachusetts. Adams and Adams, Amherst
Hitchcock E (1835) The connection between geology and the Mosaic history of creation. Biblic Repos Q Observer 5:113–38, 261–332, 439–51
Hitchcock E (1837) The historical and geological deluges compared. Am Biblic Repos 9:78–139; 10:328–74
Hitchcock E (1838) The historical and geological deluges compared. Am Biblic Repos 11:1–27
Hitchcock E (1841) Elementary geology, 2nd edn. Dayton and Saxton, New York
Kölbl-Ebert M (2002) British geology in the early 19th century: a conglomerate with a female matrix. Earth Sci Hist 21(1):3–25
Laming S, Laming D (2007) Etheldred Benett (1776–1845): the first woman geologist? In: Burek CV, Higgs B (eds) The role of women in the history of geology. The Geological Society, London, pp 247–249
Laudan R (1987) From mineralogy to geology: the foundations of a science, 1650-1830. Chicago UP, Chicago
Lawrence PJ (1972) Edward Hitchcock: the Christian geologist. Proc Amer Phil Soc 116(1):21–34
Little RD (2003) Dinosaurs, dunes, and drifting continents: the geology of the Connecticut River Valley. Earth View, Easthampton
Lonsdale K (1970) Women in science: reminiscences and reflections. Impact Sci Soc 29(1):45–59
Lowry D (1822) Conversations on mineralogy. Hunt, Philadelphia
Lyell C (1830) Principles of geology, v1. John Murray, London
Lyell C (1833) Principles of geology, v3. John Murray, London
Mason J (1991) Hertha Ayrton (1854–1923) and the admission of women into the Royal Society of London. Notes Rec R Soc Lond 45(2):201–220
Mason J (1995) The women fellows' jubilee. Notes Rec R Soc Lond 49(1):125–140
Matijasic TD (1987) Science, religion, and the fossils at Big Bone Lick. J Hist Biol 20(3):413–421
Monthly Notices of the Royal Astronomical Society (1920) Fellows elected. 80(6):563
Morrell J, Thackray A (1981) Gentleman of science: early years of the British Association for the Advancement of Science. Clarendon Press, Oxford
Newcomb S (2009) The world in a crucible: laboratory practice and geological theory at the beginning of geology. GSA, Boulder
O'Connor R (2007) Young-earth creationists in early 19th-century Britain? Towards a reassessment of 'scriptural geology,'. Hist Sci 45:357–403
O'Connor JG, Meadows AJ (1976) Specialization and professionalism in British geology. Soc Stud Sci 6(1):77–89
Penn G (1822) Comparative estimate of the mineral and mosaical geologies. Ogle, Duncan and Co., London
Prestwich GA (1901) Essays descriptive and biographical. Blackwood and Sons, Edinburgh
Rossiter MW (1971) Benjamin Silliman and the Lowell Institute: the popularization of science in ninteenth-century America. N Engl Q 44(4):602–626
Rudwick MJS (1997) Georges Cuvier, fossil bones, and geological catastrophes. Chicago UP, Chicago
Shteir AB (1996) Cultivating women, cultivating science. Johns Hopkins UP, Baltimore
Silliman B (1828) An introductory lecture delivered in the laboratory of Yale College. Howe, New Haven
Silliman B (1829) Outline of the course of geological lectures given in Yale College. Howe, New Haven
Silliman B (1839) Appendix. In: Silliman B (ed) Robert Bakewell's introduction to geology, 3rd Amer. edn. Noyes and Noyes, New Haven
Simon J (2002) Mineralogy and mineral collections in 18th-century France. Endeavour 26(4):132–136
The Athenaeum (1832) Geological sketches and glimpses of the ancient earth. No. 222: 77

The Quarterly Journal (1823) Conversations on mineralogy, with plates, engraved by Mr. and Miss Lowry, from original drawings. 14:154–160
Welsh JK (1833) Familiar lessons in mineralogy and geology, v2. Clapp and Hull, Boston
Wennerbom AJ (1999) Charles Lyell and Gideon Mantell, 1821–1852: the quest for elite status in English geology. Dissertation, University of Sydney
Whewell W (1834) On the connexion of the sciences, by Mrs. Somerville. Q Rev 51:54–68
Woodbridge WC, Willard E (1824) Universal geography, ancient and modern. Cooke and Sons, Hartford
Zornlin RM (1839) Recreations in geology. Parker, London
Zornlin RM (1852) Outlines of geology for families and schools. Parker, London

Chapter 2
The Popularization of Science

2.1 Overview

In the introduction to her most famous work *Conversations on Chemistry* (1806), Jane Marcet explained she had been frustrated upon attending public experimental lectures, finding it difficult to follow the experiments because she lacked necessary background knowledge. But she found that

> frequent opportunities having afterwards occurred of conversing with a friend on the subject of chemistry, and of repeating a variety of experiments, she became better acquainted with the principles of that science, and began to feel highly interested in its pursuit. It was then that she perceived, in attending the excellent lectures delivered at the Royal Institution, by the present Professor of Chemistry, the great advantage which her previous knowledge of the subject, slight as it was, gave her over others who had not enjoyed the same means of private instruction. (Marcet 1817: v–vi)

Marcet's epiphany should not surprise any science educator; hands-on laboratory exercises done *by* the students are clearly more effective than passive demonstrations presented *to* the class, and the seamless integration of lab experiences with a seminar or lecture offers students necessary context for the theoretical framework of the laws of nature.

In the Internet Age, members of the general public who do not have the free time or opportunity to attend a class at a local community college, or who desire additional background material concerning what they have experienced at a museum, planetarium, or public lecture (or viewed on television or read online), have countless YouTube and Khan Academy videos and TED talks offered by science media superstars from which to choose. They might even elect to read a book on the subject carefully crafted for a general, nontechnical audience (precisely what Marcet offered to her own readers). These technologies—books, museum displays, public lectures, planetarium shows, demonstrations, documentaries, MOOCs and tutorials—are examples of what is now termed the *popularization of science*.

K. Larsen, *The Women Who Popularized Geology in the 19th Century*,
DOI 10.1007/978-3-319-64952-8_2

Today the popularization of science is big business. For example, Stephen Hawking's infamously abstruse 1988 popularization *A Brief History of Time* had sold over 10 million copies by 2013 (Leane 2017). But those who create these myriad modes of science popularization have come under criticism from the scientific community for promoting, if not overt misconceptions, then oversimplifications in an attempt to stress entertainment over education (Bratton 2013). As this book will demonstrate, both the phenomenon of science popularization and criticisms of it are not novel, and indeed in some ways have changed little since the nineteenth century.

There were several different nontechnical audiences to which popularizers of science could market their works in the nineteenth century. At the least technical level were children. Despite the lack of opportunities for women at the more technical levels of science, science books for children appear to have been inclusive, aimed at both boys and girls (Fyfe 2003). The women who were these children's teachers, tutors, and mothers (often serving in several of these roles simultaneously) were also in need of accessible information on the latest discoveries in science. In addition, there was an increasing audience of working-class men who were also interested in scientific topics, but did not have the background nor disposable income to make expensive tomes written by scientists for other scientists a good fit to their needs. Therefore adult women and men often read children's science books as well, as publishers did not take advantage of this lucrative market until nearly mid-century (Fyfe 2003). This phenomenon occurred on both sides of the Atlantic, and between public lectures and demonstrations and popular-level and children's books, the popularization of science in both Britain and America reached nearly a fever pitch by 1850 (Kohlstedt 1990).

While there are a number of similarities between the movement to popularize science in the nineteenth and twentieth centuries, there are two significant differences, the first being one of the central reasons for bringing science to nonscientists in the nineteenth century, and the other the demographics of the authors to whom the responsibility for writing these works fell. In the nineteenth century science was widely utilized as vehicle through which to teach piety and proper behavior, especially to children and young ladies. The lessons of the natural world were intended not only to educate, but to inspire, in a religious sense, as seeing God's hand reflected in his handiwork was often considered more important that knowing the science itself (as described by Edward Hitchcock [Sect. 1.3]) (Fyfe 2008).

What is of interest is that many of these early superstars of science popularization in the nineteenth century were not practicing scientists at all, but instead teachers, tutors, and amateur enthusiasts, many of whom were self-taught in the very subject they attempted to explain to a general audience. And, not coincidentally, they were female. While many of their names are nearly forgotten today, a handful persist in the history of science, held up as curiosities, exceptions to the rule of the sexist scientific establishment of their century. One such example is Mary Somerville and her wildly popular *On the Connexion of the Physical Sciences* (1834). But Somerville also wrote a highly influential book on Physical geography that deserves far more analysis by historians of science than it has received.[1] For this reason, she is the subject of Sect. 9.3.

Admittedly the current lack of name recognition of these women is not merely due to their gender. Until recent decades, the popularization of science in general was not a topic of wide interest to historians of science due to the relative unimportance with which popularization was considered (Cooter and Pumfrey 1994). According to the positivist diffusion model of science that held sway over much of the twentieth century, scientists held the privileged active position of being the sole producers of knowledge. The popularization of science was considered to be a low-status activity that was not worth the time of scientists themselves. This caste system in communicating science mandated that scientists spoke to each other through peer-reviewed journals and left the popularization of science to journalists and those without the academic credentials to be practicing scientists (Hilgartner 1990; Whitley 1985).[2] While some well-known scientists took an active role in the popularization of science in the early twentieth century, for example astronomer Sir James Jeans in Britain, by the late twentieth century those scientists who chose to speak with the general public found themselves *Saganized*, a pejorative term referring to Pulitzer Prize winning astronomer Carl Sagan, who suffered backlash from the scientific community for his media efforts to bring science to the average person. As with other aspects of the popularization of science, this is also not a new phenomenon. In his *Discourses: Biological and Geological Essays* (1898), Thomas H. Huxley warned that "the popularization of science, whether by lecture or essay, has its drawbacks. Success in this department has its perils for those who succeed. The 'people who fail' take their revenge… by ignoring all the rest of a man's work and glibly labeling him a mere popularizer" (vii–viii).

While the popularization of science has become less openly derided in recent decades (due, in part, to the best-selling books of physicist Stephen Hawking and paleontologist Stephen Jay Gould, among others), scientists who engage in popularization can still face prejudice from the scientific community when applying for grants or when applying for promotion or tenure. The logic is that if one is devoting time to popularization, one's true scientific work (i.e. research) must be suffering in the process (Ellison 2004). Such prejudice is partly behind efforts to avoid the term *popularization* and replace it with *expository science*, a term with less political baggage (Cooter and Pumphrey 1994).

At the heart of this prejudice is a patent misunderstanding of the nature of popularization, of conflating it with merely simplifying—of watering-down—the pure science in order to make it palatable to the general public (Gavroglu 2012). Increasingly both scientists and historians of science have come to understand that the effective popularization of science requires two skill sets rather than one; not only does the author have to possess an understanding of the science at hand, but the

[1] For more information on the popularization of science in general in the nineteenth century, see Shtier (1996), Gates (1998), Lightman (2007), and O'Connor (2013).

[2] For an overview of the politics surrounding the popularization of science, see Bensaude-Vincent (2009).

effective communication skills to *translate* the science into a more common vernacular. This is no mere simplification. There is therefore not only room for, but a need for, both those who create the scientific knowledge and those who more widely disseminate it beyond the technical discipline.

As both the early nineteenth century and twenty-first century scientific communities have shown, there is sufficient space for researchers, educators, and authors in science; however, as discussed in Sect. 1.4, this was a highly gendered space in the nineteenth century, with women being relegated to clearly limited and secondary roles. In particular, the popular-level scientific literature of the 1700s and 1800s also reflected a gender segregation by discipline. Certain sciences and subspecialties were considered appropriate for women to write about, while others were deemed the province of men. Perhaps the premier example of a female-friendly science was botany, while physics and chemistry were largely considered part of the realm of men. However, there have always been individual women who chafed against the hegemony, and refused to be relegated to the sidelines. Two important examples are Mary Somerville, as seen in her aforementioned bestseller *On the Connexion of the Physical Sciences,* and Jane Marcet's *Conversations on Chemistry*. In particular, Marcet's work demonstrated the pedagogical power of uniting a nontechnical explanation of science with opportunities to experience science through hands-on experimentation: the ability to *do* science, even at a simplistic level. This marriage of literature and science led to a revolution of science-based literature largely written *by* women *for* women (and their children) called the *familiar format*.

2.2 Defining the Familiar Format in Science Writing

As previously noted, a market for popular-level books on science (especially aimed at an audience of children and women) opened up in the late eighteenth century. But this new audience required an equally novel style of writing, meant to effectively transmit scientific knowledge to those with a minimal background in science. This methodology soon became known as the *familiar format*, or alternately the conversational style (although the conversational aspect was merely one of the defining characteristics of this type of writing). While a number of scholars have described this literary style (e.g. Shteir 1996; Gates 1998; Fyfe 2003; Lightman 2007; Larsen 2014) as well as offered specific examples of it (most often in botanical texts, where it achieved its widest usage), it is difficult to find a detailed template for what constituted a true work in the familiar format, as well as an analysis of exactly how much latitude the authors had in branching out from what soon became a literary trope. Such a detailed analysis forms the backbone for much of the rest of this volume.

While there was certainly considerable variety in the works themselves, there are seven common characteristics that appear to be central to the genre's characterization:

1. The audience is presumed to be laypersons (including children, women, and working class men), and as such, the works are openly adapted by the author for their audience. This includes simplifications (often including a lack of mathematics) or some other intentional adaptation to the background and abilities of the audience. The exact type of adaptation is usually directly addressed by the author, often in the preface or introduction, and may also explain to the audience what is expected on their side. As will be shown, this does not necessarily include the watering down of scientific terminology (although some terminology may be relegated to an appendix attached to the work proper);
2. The literary style is most often that of a fictional dialogue or letters, rather than a lecture written in third-person. Note that there is a wide latitude found in the depth of the characters; in some cases the individual characters are well fleshed-out and have distinctive personalities and backgrounds, giving the work a rather novelistic feel. Also included in this literary style are one-sided conversations where the author appears to be speaking *with*, rather than formally lecturing *to*, the anonymous audience/reader (often made clear through the use of more personal first-person rather than formal third-person pronouns), as well as catechisms;
3. The authority figure who imparts the scientific knowledge to the other characters (and hence the audience) is either a woman (a mother or tutor) or, less frequently, a pair of parents. In the latter case, the mother takes an active role in describing science, although this role may be obviously both secondary and subservient to the role of the father;
4. The work is set in a domestic setting, usually either a home or garden, or more generally outdoors, rather than a lecture hall or formal classroom. This affords ample opportunities for experimentation and observation in a comfortable and informal manner;
5. The characters are actively engaged with the natural world as part of the learning process. They make observations, gather samples, and often gain hands-on experience with technology (such as microscopes or goniometers). If the work is written in the form of a catechism or one-sided author conversation with the reader, the reader/student is clearly directed to engage in observation and hands-on learning by the author/teacher;
6. Everyday examples and common metaphors are often used to describe the science. Experiments may be done using household items (as seen in Fig. 2.1) and the science of these items may likewise be explained;
7. Morality and religiosity are often clearly and repeatedly reinforced by the author/authority figure. The benefit of scientific study (especially its ability to reinforce ethical behavior and religious piety) is normally stressed.

Between 1780 and 1840, this format was adopted by a number of writers of science works for children and women, and afforded women a gender-appropriate space within specifically science writing and education, but more broadly within the scientific endeavor. Fyfe (2008) notes that the genre reached its peak early in the nineteenth century, with an average of two scientific *Conversations* published per

Fig. 2.1 Spinning a clay ball on a stick to represent the oblate spinning earth. Reproduced from Delia W. Godding (1847) *First Lessons in Geology*

year in the 1820s. The title page of a representative example is shown in Fig. 2.2. The use of letters or conversations to communicate science was not a novel idea, having been previously used by such illustrious (male) scientists as Galileo Galilei and Robert Boyle (Shteir 1996; Gates 1998).

The appearance of science in a domestic setting not only echoes the argument of feminist writer Mary Wollstonecraft (1759–1797) that education should reflect the everyday experience of each individual (Shteir 1996), but also the architecture of the day. In Europe, the availability of salons and drawing rooms offered a natural space in which myriad topics of interest, including politics and science, could be discussed in comfort, leading to the salon tradition that played a central role in the dissemination of scientific knowledge to women of the upper socioeconomic classes. In the United States, most middle-class homes had parlors, rooms where visitors were welcomed, conversations held, and books housed. For example, Yale geologist Benjamin Silliman recalled reciting lessons with his siblings in his family's parlor, demonstrating the use of the parlor as an educational space (Kohlstedt 1990). Not only were these parlors furnished with tables, chairs, and bookcases, but given their ceremonial purpose as a welcoming space for guests they would have also been the location of the family's natural history cabinet. For these reasons as

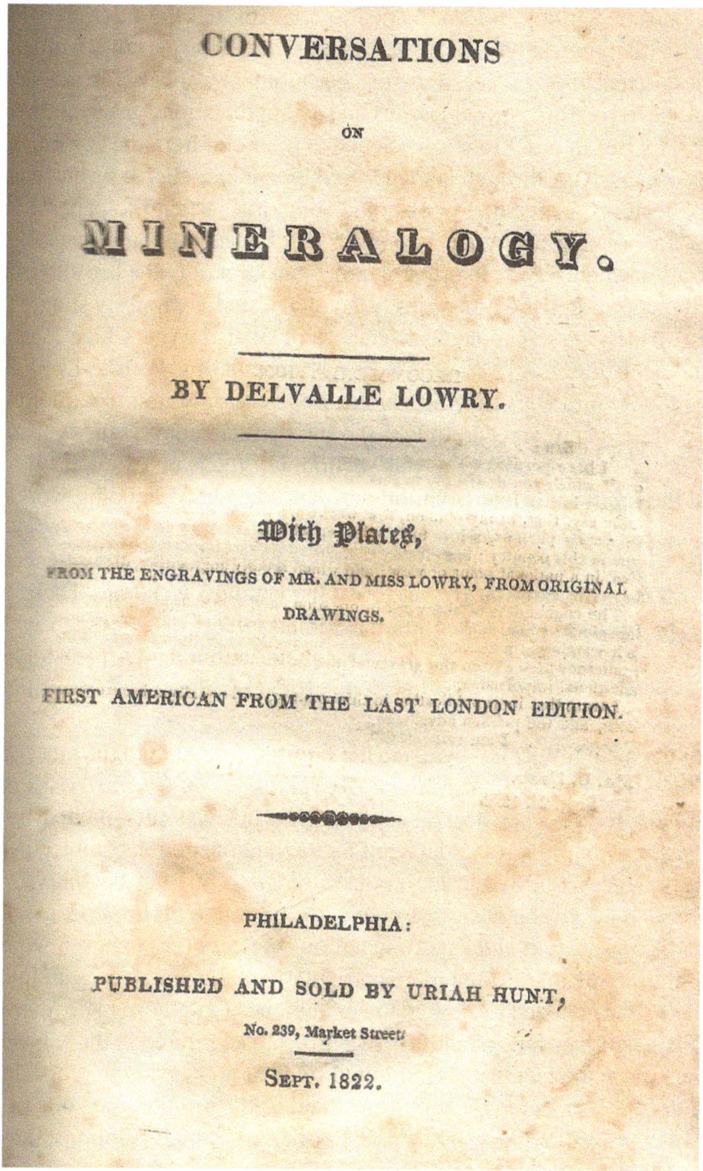

CONVERSATIONS

ON

MINERALOGY.

BY DELVALLE LOWRY.

With Plates,

FROM THE ENGRAVINGS OF MR. AND MISS LOWRY, FROM ORIGINAL DRAWINGS.

FIRST AMERICAN FROM THE LAST LONDON EDITION.

PHILADELPHIA:

PUBLISHED AND SOLD BY URIAH HUNT,

No. 239, Market Street.

SEPT. 1822.

Fig. 2.2 Title page. Reproduced from Delvalle Lowry (1822) *Conversations on Mineralogy*, Amer. Ed

well, the parlor would have afforded a convenient space for scientific investigations, such as using microscopes and magnifying glasses to examine the fossils and minerals that could have been collected on everyday walks (depending on the geographic location).

This domestic setting, in turn, helped to confirm that the matriarchal transmission of scientific knowledge was a societal expectation as part of a woman's household duties. In teaching science, these teacher/mothers were also afforded additional opportunities to reinforce moral lessons and an appreciation for scriptural teachings (Shteir 1997). But in order to teach science to their children, these women first had to educate themselves, through being tutored by a male relative or husband, attending public lectures, or reading books (and sometimes scientific papers) written by scientists. In order to aid other women in this endeavor, some turned to writing popularizations themselves. In order to understand not only the breadth of the genre but how it incorporated the properties listed above, and to properly situate the familiar format writings in the geological sciences of British writers Jane Marcet, Maria Hack, and Delvalle Lowry, and Americans Jane Kilby Welsh, Delia Woodruff Godding, and Almira Hart Lincoln Phelps, it is necessary first to explore the wider landscape of the genre and carefully differentiate mere conversations from the true familiar format. The remainder of this section will focus on examples of conversations and the familiar format from the physical sciences (including works by male writers in geology), while the following section will specifically focus on early influential women writers in the genre, particularly in botany, the science in which female writers of the familiar format found the warmest welcome. These works are vital to explore, as not only were they read and cited by the women whose work forms the core of Chaps. 3–8, but provided them with a ready-made audience that had demonstrated its eagerness for popular-level science works written in this format.

One of the most influential works in the familiar format (and the most commonly cited work in the genre) is Jane Marcet' *Conversations on Chemistry*. As previously noted, in her preface Marcet explains that the work was precipitated by her own experiences in attempting to understand public demonstrations and lectures. Based on her experiences, she wrote the book in order to instruct beginners, and most especially women, in the basics of chemistry by mimicking the form of her own instruction: conversation and experimentation. Marcet notes that as a woman "venturing to offer to the public, and more particularly to the female sex, an Introduction to Chemistry," she needed to explain to her audience (and those male reviewers who might look askance at such a bold work) her own background in the subject (Marcet 1817: v). Despite the fact that she admitted some apprehension in proposing a work which might be considered by some "unsuited to the ordinary pursuits of her sex," she "felt encouraged by the establishment of those public institutions, open to both sexes, for the dissemination of philosophical knowledge, which clearly prove that the general opinion no longer excludes women from an acquaintance with the elements of science" (Marcet 1817: ix). However, it must be noted that there is a significant difference between allowing women to attend public lectures in the physical sciences and accepting them as authors of works in these historically male sciences, especially a work which was originally published anonymously. To Marcet's undoubted surprise, *Conversations on Chemistry* was widely-acclaimed, and enjoyed numerous editions in both England and the United States. The work's most famous reader was scientist Michael Faraday, who received his first introduction to electrochemistry through its pages while he was a young bookbinder. In his later years he often gave credit to Marcet for sparking his interest in chemistry.

As she begins her book, Marcet is fully aware of the potential anxieties of her audience, and explains that

> it was natural to infer, that familiar conversation was, in studies of this kind, a most useful auxiliary source of information; and more especially to the female sex, whose education is seldom calculated to prepare their minds for abstract ideas, or scientific language. As, however, there are but few women who have access to this mode of instruction; and as the author was not acquainted with any book that could prove a substitute for it, she thought that it might be useful for beginners, as well as satisfactory to herself, to trace the steps by which she had acquired her little stock of chemical knowledge, and to record, in the form of dialogue, those ideas which she had first derived from conversation. (Marcet 1817: vi–vii)

Given the wide popularity of her work, the familiar format proved a poweful pedagogical tool when wielded by her hand.

Perhaps the breadth of Marcet's influence on science writing is best seen in a short-lived expansion of the familiar format that her works sparked. Even Charles Lyell himself considered writing a geological conversation, but thought against it, given the limitations of the style. In an 1828 letter to Gideon Algernon Mantell, Lyell explained

> I at first intended to write 'Conversations on Geology:' it is what no doubt the booksellers, and therefore the greatest number of readers, are desirous of. My reason for abandoning this form was simply this; that I found I should not do it at all, without taking more pains than such a form would do justice to. Besides, I felt that in a subject where so much is to be reformed and struck out anew, and where one obtains new ideas and theories in the progress of one's task, where you have to controvert, and to invent an argumentation—work is required, and one like the 'Conversations on Chemistry' and others would not do.... But finally, I thought, that when I had made up my own mind and opinions in producing another kind of book, I might then construct conversations from it. In the meantime there is a cry among the publishers for an elementary work, and I much wish you would supply it. Anything from you would be useful.... (Lyell 1881: 177)

Lyell instead wrote his classic (and traditional lecture-style) text *Principles of Geology* (in three volumes published between 1830 and 1833) and *Elements in Geology* (1838) and never wrote a conversation-based geology book.

Not all so-called familiar books are written in a truly familiar format (using the seven points previously identified for the genre). For example, British mineralogist John Mawe's *Familiar Lessons on Mineralogy and Geology* (1819) is written in straightforward lecture style. Despite this it proved so popular that by 1830 it was already in its 12th edition.

Mantell eventually wrote several popular-level works in geology, most notably *The Wonders of Geology; or, a Familiar Exposition of Geological Phenomena* (1838) and *The Medals of Creation; or First Lessons in Geology and in the Study of Organic Remains* (1844). Neither of these works used the conversation style. A book entitled *Conversations on Geology* did appear in 1828, with the cumbersome subtitle "Comprising a familiar explanation of the Huttonian and Wernerian systems; the Mosaic geology as explained by Mr. Granville Penn; and the late discoveries of Professor Buckland, Humboldt, Dr. Macculloch, and others." The book, published anonymously but now attributed to Scottish naturalist James Rennie (1787–1867), was considered such an enthusiastic promotion of Penn's Mosaic geology that through the mid-twentieth century many scholars attributed it to Penn himself (O'Connor 2007).

In Rennie's work a fictional Mrs. R explains the various systems of geology to her inquisitive children Edward and Christina. As is common in the familiar format, Rennie offers to his audience that he has two reasons for writing this book; to "describe the various Theories of the Earth and Systems of Geology which have, from time to time, prevailed, and have brought forward facts and reasonings either in support or contradiction of them" and to make the reader "enabled to 'Look through Nature up to Nature's God'" (Rennie 1828: vi). He then includes five pages of quotes from William Buckland supporting the view that the earth has been designed by God for the benefit of humanity. The first conversation is on "Theories of the Earth" and starts with Edward asking his mother about seashells found "in the heart of solid rocks, and far inland" (Rennie 1828: 1). She explains that "The history of the shells, my dear, and many other things no less wonderful, is contained in the science called GEOLOGY, which treats of the first appearance of rocks, mountains, valleys, lakes, and rivers; and the changes they have undergone, from the Creation and the Deluge, till the present time" (Rennie 1828: 3). Note that the Biblical chronology is assumed to be factual from the start of the text. Edward is afraid that he won't be able to understand the science, so he asks if there is an "easy book on Geology, like the delightful 'Conversations on Chemistry,' which I could read without being puzzled?" (Rennie 1828: 4). His mother explains that geology is too young a science for such a book to exist, but she is willing to try and teach him without a book, and the conversation unfolds.

Religious and moral lessons play a central role in this work, as well as the role of geology in teaching both. For example, Mrs. R opines that "Geology is, perhaps, better fitted for our limited comprehensions than astronomy; for it is more within our reach to examine the structure and formation of mountains, than that of the sun or the stars…. we are not so overpowered by sublimity as in the contemplation of astronomy; we can think more calml [sic] and reason more at ease; and we can trace the finger of God more visibly, perhaps, because more nearly" (Rennie 1828: 8–9). In particular, after a lengthy introduction to the "mineral geology" of Werner and Hutton, Rennie (1828: 45) prefaces his discussion of Penn's model by calling it "no less ingenious than probable, and will give you much more sublime views of the creation than are to be found even in the inspired poem of Milton; and that is saying a great deal."[3] The children learn through the observation of rock samples from Mrs. R's cabinet, but given the premise of the book (the theories of the history of the earth), there are no experiments conducted. This is a significant limitation of this work, and therefore it is difficult to classify this as a true example of the familiar format.

Another example of a work that does not fulfill the definition of the familiar format is the six-volume work *Scientific Dialogues* (1800–5), written by Reverend

[3] For a detailed analysis of Rennie's style, see O'Connor (2007).

Jeremiah Joyce (1763–1816), the former tutor of the Earl of Stanhope's sons. Although written as a series of conversations, the teacher figures are both male (a male tutor and a father) while the students include two boys, Charles and James (named for the Earl of Stanhope's sons), and a girl, Emma. Joyce's work was extremely popular, enjoying seven editions from 1800 through 1892, one of its most famous readers being John Stuart Mill (Issitt 2002). In keeping with the multiple audiences targeted by such works, Joyce explained that

> the whole work will be found a complete compendium of natural and experimental philosophy, not only adapted to the understanding of young people, but well calculated also to convey that kind of familiar instruction which is absolutely necessary before a person can attend public lectures in these branches of science with advantage. (Joyce 1852: i)

The work opens with a discussion between siblings, Charles and Emma, and their Father that gives direct credit to another famous series of children's stories:

> Charles. Father you told sister Emma and me, that, after we had finished reading the "*Evenings at Home*," you would explain to us some of the principles of natural philosophy: will you begin this morning?
>
> Father. Yes, I am quite at leisure…. These, I trust, will lead you insensibly to admit the wisdom and goodness by means of which the whole system of the universe is constructed and supported. (Joyce 1852: 1)

Note that the religious undercurrent of the work is present from the beginning, and is repeated throughout the work.

Emma is hesitant to begin the lesson, and, as is common in such works, voices a concern that she is too young to understand or appreciate science. She asks "can philosophy be comprehended by children so young as we are? I thought that it had been the business of men, and of old men too." Her father assures her that "Philosophy is a word which in its original sense signifies only a love or desire of wisdom; and you will not allow that you and your brother are too young to wish for knowledge" (Joyce 1852: 1). Joyce sets the tone for his approach to popularization in the same discussion, as Charles offers "But in some books of natural philosophy, which I have occasionally looked into, a number of new and uncommon words have perplexed me; I have also seen references to figures, by means of large letters and small, the use of which I did not comprehend." His father understands his son's concerns, offering

> It is frequently a dangerous practice for young minds to dip into subjects before they are prepared, by some previous knowledge, to enter upon them; since it may create a distaste for the most interesting topics….The same sort of disgust will naturally be felt by persons who should attempt to read works of science before the leading terms are explained and understood. (Joyce 1852: 1–2)

Joyce then begins his lengthy examination of this topic by introducing the children to the concept of measuring and labeling angles, demystifying the topic at the start by deconstructing the very diagrams that Charles had voiced concern about.

In an interesting compromise experiment, English geology lecturer and mineralogical surveyor Robert Bakewell (1768–1843) attached to his lecture style *An Introduction to Mineralogy* a "Series of Conversations Explaining the Principles of the Science, and the Elements of Crystallography" (Bakewell 1819). In his introduction, Bakewell explains that "In the German method of describing minerals, the meaning is sometimes almost buried under a heap of terms" (1819: vi). As a result, he "adopted the form of dialogue in some of the introductory chapters of the present volume, that the student may become more familiar with the subject, by seeing it presented under different points of view, and may receive explanations and answers to certain queries and objections, that will naturally occur to his mind" (Bakewell 1819: vii). The conversation begins between two men, B, and F, who became separated during a hike because F was intrigued by sparkling crystals that he hoped were diamonds. B explains the true nature of the quartz crystals, which leads to a discussion of mineral properties. The style is very curt, a simple back and forth with few clues regarding the personalities of the two men or their relationship. This work in particular demonstrates the important difference between a mere conversation and a fully initiated familiar format work, especially in its lack of female authority figures.

One of the earliest popular examples of the true familiar format was *The Newtonian System of Philosophy; explained by familiar objects, in an entertaining manner, for the use of young ladies and gentlemen,* by Tom Telescope, the pen name of publisher John Newberry (1713–67), who played a pivotal role in the 1740s creation of the genre of books specifically for children (Fyfe 2003). Despite the fact that the subtitle claims that the work is appropriate for child readers of both genders, the active characters in the books are young boys, including the narrator, Tom Telescope, and several adult males. Several adult women (the wives of the male characters) do take part in the conversation, but are relegated to largely passive roles, something that is more prevalent in familiar format works written by men.

In this work, a series of schoolboys gathers over the holidays to play games at the home of Lady and Lord Twilight. But the young gentleman Tom takes the moral high ground, warning that

> Playing at cards for money, says he, is so nearly allied to covetousness and cheating, that I abhor it.... I should prefer those which not only divert the mind, but improve the understanding: and such are many of the diversions at the school where I am placed. We often play at sham Orations, comical Disputes, measuring the Land and Houses, taking the Heights and Distance of Mountains and Steeples, solving Problems and Paradoxes on Orreries, Globes and Maps, and sometimes at Natural Philosophy.... (Telescope 1803: 2–3)

Mrs. Twilight is intrigued by this possibility of taking an active part in science (and in science as a type of play), and leads the children

> to Mr. Setstar's, that they might have the use of proper instruments. As Mr. Setstar was engaged in company, Lady Twilight, though nearly related to him, would not disturb him, but led them through the saloon [sic] into a private parlour, where our little Philosopher, at the request of her Ladyship, immediately opened the Lecture, without making idle excuses, or waiting for farther solicitations.... (Telescope 1803: 3)

Note the clearly domestic setting of the story in the parlor of the house, as described above. The remainder of the book is a set of conversations on various aspects of science, involving adults and children, always with an eye towards both experimentation and observation, but connecting the scientific concepts to everyday life. For example, in a discussion of minerals Tom explains "That reflecting telescope, this gold watch, and Lady Caroline's diamond ear-rings, were all dug out of the earth; at least the materials were there found, of which these things are composed" (Telescope 1803: 88).

A second commercially popular early example of the familiar format was the work referenced in Joyce's book, a series of stories entitled *Evenings at Home; or the Juvenile Budget Opened.* This collection was originally published in six volumes between 1792–6 (and reissued in various editions through the 1850s) by John Aikin and his sister Mrs. Anna Laetitia Barbauld.[4] These stories involved the Fairbourne family who lived in the village of Beechgrove and entertained (and were entertained by) many visitors, always with an eye to improved morality. In order to contribute to the education of the Fairbourne children, these visitors would

> frequently produce a fable, a story, or dialogue, adapted to the age and understanding of the young people.... and when the pieces were once read over, they were carefully deposited by Mrs. Fairborne in a box, of which she kept the key. None of these were allowed to be taken out again till all the children were assembled in the holidays. It was then made one of the evening amusements of the family to *rummage the budget*, as their phrase was. (Aikin and Barbauld 1819: 2)

Not only do both men and women, girls and boys, take part in the scientific investigations in this domestic setting, but as in the case of Tom Telescope's investigations, hands-on experiential learning is stressed. For example, on the 21st evening, in a story entitled "On Earths and Stones," Harry and George discuss with their tutor different types of "calcareous earth," including marble, chalk, and limestone. The tutor directs George to pour vinegar on a "piece of chalk or an oyster-shell" on his own to see the effects of such minerals dissolving in acid (Aiken and Barbauld 1805: 8). *Evenings at Home* clearly meets the definition of a familiar format work, and sets an early standard that other works aspired to achieve.

Having surveyed some of the various examples of both the successful and incomplete application of the familiar format to writings in the physical sciences (and mainly by male writers), our attention now turns to early female writers in the genre. Both the works previously described and those highlighted in the following section largely formed the corpus of works to which those by Marcet, Hack, Lowry, Welsh, Godding, and Phelps would have been compared by both their readers and critics, as well as influential works from which these writers would have drawn their inspiration. These writers also demonstrated by example that such writing was, indeed, accepted by their society as proper for a lady, and that in some cases, women could support not only themselves, but their parents and/or children through science writing.

---

[4] For a detailed analysis of the works, see Levy (2006).

2.3 Early Women Authors of the Familiar Format

In her 1798 work *Reflections on the Present Condition of the Female Sex; with suggestions for its improvement,* Priscilla Wakefield opines "There are many branches of science, as well as useful occupations, in which women may employ their time and their talents, beneficially to themselves and to the community, without destroying the peculiar characteristic of their sex, or exceeding the most exact limits of modesty and decorum" (Wakefield 1798: 8–9). The writing of botany books, especially in the familiar format, for children and their mothers was one such occupation in the late eighteenth and early nineteenth century. As previously described, the genre incorporated a number of literary styles, from simple catechisms to letters and fully developed conversations. Shteir (1990) argues that the catechism was the least widely used of these formats by women science writers, perhaps because it was less personal and in its strictest form stressed memorization over contemplation. However, as will be shown, in the hands of some women writers even this normally restrictive form could serve the purpose of the familiar format. Letters were also less widely used than the conversation, although as Sam George (2011) notes, the epistolary form has a history of use in the writing of books meant to focus a young woman's attention on proper ladylike behavior and societal norms (for example the conduct or advice book). It is now instructive to survey a sample of familiar format works written by women during the late eighteenth and early nineteenth centuries in order to illustrate both the common points of the content of the genre as well as the variety of ways in which it was executed by women writers.

Botany was the subject of the vast majority of familiar format works written by women between 1780 and 1830 for two important reasons. Firstly, the physical sciences were seen as a male bastion, and there were therefore far fewer women writers in these fields. Secondly, as noted by Lady Charlotte Murray (1754–1808) in her work *The British Garden* (1799), "The expensive apparatus of the Observatory, and the labours of Chemistry, confine the science of Astronomy, and the study of Minerals to a few…. But the study of Botany … is open to almost every curious mind…." (Murray 1808: vi). One of the first women to take advantage of this role for science minded women was Maria Elizabeth Jacson (alternatively Jackson) (1755–1829). As was typical of the day, she anonymously published her works, *Botanical Dialogues: Between Hortensia and her Four Children* (1797), *Botanical Lectures* (1804), *Sketches of Physiology of Vegetable Life* (1811), and *Florist's Manual, or Hints for the Construction of a Gay Flower-Garden, with Directions for Preventing the Depredations of Insects* (1816). Jacson was the daughter of Midlands rector Simon Jacson, who had received a clergyman's education at St. John's College, Cambridge. She and her older sister, Frances Margaretta (1754–1842), a novelist who, like Maria, never married, lived with their widowed father until his death in 1808 (Shteir 1990). Maria and Frances afterwards lived in Derbyshire with their elder brother Roger (1753–1826), but his second wife made the situation untenable, and they later settled in a manor house that had belonged to a maternal aunt (Percy 1997).

Jacson's first book, *Botanical Dialogues*, was, as its subtitle explains, a conversation between Hortensia and her children, Charles, Harriet, Juliette and Henry. Although the book was only attributed to *a Lady*, it contained within it a letter of support signed by two well-known figures, Erasmus Darwin (Charles Darwin's grandfather) and Brooke Boothby. It was their opinion of the text that "not only the youth of both sexes, but the adults also, will be much indebted to your ingenious labours" (Jacson 1797: iv). Jacson likely met Erasmus Darwin—well known doctor, scientist, and author of the long poem *The Botanic Garden*, as well as translator of Linnaeus, principal member of the Botanical Society of Litchfield, and enthusiastic proselytizer of Linnaean botany—through her cousin (on her mom's side) Sir Brooke Boothby (Shteir 1990). Darwin also recommended *Botanical Dialogues* as an effective textbook in his own *Plan for the Conduct of Female Education* (1797), a work meant for his daughters Susan and Mary Parker's girls' boarding school. Shteir (1990) interprets diary evidence to suggest that Jacson may have herself been friends with the Parker sisters.

In terms of its literary style, Jacson's dialogue is written in a rather script-like form, as shown:

> *Hortensia*.... We will begin our lectures this morning. I have promised Henry and Juliette that they shall be of our parties, they are never idlers either at lessons or play, and will, I dare say, find both amusement and instruction from the study.
>
> *Henry*. We will be very attentive.
>
> *Juliette*. I long to know the names of all those pretty things, that we find when we pull a flower in pieces.
>
> *Hortensia*. I am a little afraid, lest the hard names should be too difficult for my youngest pupils, however I will endeavour to make them easy.... (Jacson 1797: 3)

Note the acknowledgment of difficult scientific names, and the intention to simplify the topic for her young readers. However, Jacson is clear to delineate the limits to which she is willing to go in terms of simplifying the science:

> *Harriet*... sometimes I feel a little afraid of being found dull; and I think I have heard of botanical books written for ladies, which make all the hard words easy.
>
> *Hortensia*. There are some books, which pretend to do it; but the scientific terms are still to be learnt, and when learnt, in the language of those books, you cannot converse with a Linnaean botanist; they may make you a partial, but cannot make you an [sic] universal botanist.... (Jacson 1797: 71)

As previously noted, simplification does not necessarily mean the watering down of scientific terms.

Hortensia engages her children with the science through hands-on activities. For example, Jacson includes descriptions of observations of plants and their parts, and how to classify them, in addition to utilizing a microscope to examine them in more detail. Interestingly, when Hortensia introduces Charles and Henry to a more obscure type of plant, she explains to her sons "This tribe of plants not having been much attended to leaves to modern botanists an ample field of discovery, and I flatter myself it is reserved for you, Charles and Henry, to distinguish yourselves in it." When Charles asks why this knowledge is not also for his sisters, his mother explains

> I do not doubt their abilities, and would have them as thoroughly informed upon the subjects that they study, as I wish you to be, but to avoid obtruding their knowledge upon the public. The world have agreed to condemn women to the exercise of their fingers, in preference to that of their heads; and a woman rarely does herself credit by coming forward as a literary character. The world improves, and consequently female education. Some years ago a lady was ashamed to spell with accuracy; happily the matter is now reversed, and the time will come, when it must be granted, that by improving our understandings, we enlarge our view of things in general; and thence are better qualified for the exercise of those domestic occupations, which we ought never to lose sight of, as our brightest ornament, when properly fulfilled. At this time information in a woman, beyond a certain degree, distinguishes her above her companions, and like all other distinctions is liable to lead her into a vain display, of what she hopes will gain her admiration…. (Jacson 1797: 238–9)

Shteir (1990) highlights the uncharacteristically sexist nature of this passage, and explains it as reflecting society's expectations for a proper lady of the time. A lady should be demure, modest, and certainly never provide intellectual competition for her husband.

There is also the obligatory connection between God and the natural world. As an example, Hortensia offers to her children concerning grains "We cannot contemplate the fruits of the earth, which are so bountifully bestowed on all climates, and the faculties with which man is endued to discover their uses, but we must adore in silent and grateful praise, the beneficent Creator of all things…." (Jacson 1797: 303). Unfortunately the book was not a commercial success and did not receive a second edition. Jacson afterwards abandoned the familiar format for a more straightforward and impersonal textbook style of writing in her later works.

Priscilla Wakefield (1750–1832) wrote over a dozen works, many in the familiar format. She was the eldest of six children in the Quaker household of Daniel Bell (1726–1802) and Catherine Barclay (1727–84), and in the tradition of her family's faith, she was educated at home by her mother. In turn she later played a role in the education of her own children and grandchildren, following her marriage to Edward Wakefield (1750–1826), a merchant (Leach 2006). Wakefield's Quaker faith and upbringing deeply shaped her convictions about the role of education for both boys and girls, including a basic understanding of science. This is reflected in her preface to *Mental Improvement, or the Beauties and Wonders of Nature and Art, in a series of instructive conversations* (1797). Wakefield believes there are "four methods of attaining knowledge. Observation, reading, conversation, and meditation. The first lies within the compass even of children, and from the early dawn of reason, they should be accustomed to observe every thing with attention, that falls under their notice" (Wakefield 1799: iii). This affords the "judicious instructor" with countless opportunities for developing "a lesson among, those objects, that are termed common or insignificant," but as she laments, these opportunities are not often taken advantage of, given the observed "ignorance, not of children only, but sometimes of youth, who, although they have attained a considerable degree of classical learning, are unacquainted either with the materials of those things they daily use, or the methods of manufacturing them" (Wakefield 1799: iii). In response, Wakefield takes care to integrate everyday items and experiences into her own instructional works, and encourages the direct, careful observation of these items.

Mental Improvement features the Harcourt family, two parents and their four children, ages nine to sixteen, who engage in evening lessons (many of which directly involving science) in their home. The children invite a friend, Augusta, to take part in their familiar lessons, taking pity on her because her mother is dead, an important moral lesson taught in the first pages of the book. Both parents take an active role in the instruction, on topics as varied as whales, telescopes, salt, honey, and wool. One evening Augusta apologizes for arriving too early for the lesson. Mr. Harcourt understands her eagerness to learn, and suggests that it is because of the pleasantness of the lessons offered. Wakefield then explains, through Mr. Harcourt, her personal pedagogical philosophy:

> Instructions should always be rendered agreeable, in order to be beneficial to those that are to learn. The skill of a preceptor consists in gaining the affections of his pupils, and conveying knowledge in so gradual and clear a manner, as to adapt it to the strength of the young student's capacity. Many a poor child has been disgusted with books and learning, by the heavy laborious talks that have been given him to learn by heart, before he was capable of understanding them. (Wakefield 1799: 121)

Here we see an accusation still lobbed today at traditional passive modes of education, that the natural curiosity of children is dampened or even extinguished through rote memorization and boring lectures. To avoid this, Wakefield not only engages children in active learning, but introduces the content slowly and clearly (her mode of adaptation). Wakefield therefore fulfills all seven benchmarks of a work in the familiar format.

Wakefield wrote other works in this genre as well. For example, *Domestic Recreation; or Dialogues Illustrative of Natural and Scientific Subjects* (1805) features dialogues between Mrs. Dimsdale and her two daughters, Lucy and Emily. The family actively explores the natural world through personal observation. As the work begins, Lucy brings her mother an insect she has found and points out its "horns." Her mother corrects her, naming them "antennae, or feelers" and proceeds to explain their presumed purpose (Wakefield 1805: 1). Emily offers, from her own personal experience, that a butterfly's antennae are different, and Mrs. Dimsdale directs Emily to "take this key, and in my tortoiseshell cabinet you will find a drawer, in which there are the parts of many insects dried: bring it, and we will examine them" (Wakefield 1805: 2–3). Note that the Dimsdale family, like many of the families who would purchase Wakefield's books, owns a natural history cabinet that can be utilized in the lessons.

Wakefield's familiar format work in botany was *An Introduction to Botany, in a Series of Familiar Letters* (1796), framed as letters sent from one sister (Felicia) to another (Constance) during a period of separation. To pass the time and bridge the miles, Felicia explains that she will summarize her botany lessons for Constance in her letters, lessons arranged by their mother as they "will be beneficial to my health, as well as agreeable, by exciting me to use more air and exercise than I should do, without such a motive; because books should not be depended upon alone, recourse must be had to the natural specimens growing in fields and gardens" (Wakefield 1807: 2-3). As the letters are all from Felicia to Constance, the book reads more like a lecture than her conversational works, although they still include Felicia's hands-on observations of actual plants. Wakefield's intention for penning these letters mirrors the dual goals of education and religious inspiration so common in this genre:

> The design of the following Introduction to Botany, is to cultivate a taste in young persons for the study of nature, which is the most familiar means of introducing suitable ideas of the attributes of the Divine Being, by exemplifying them in the order and harmony of the visible creation.... The structure of a feather or a flower is more likely to impress their minds with a just notion of Infinite Power and Wisdom, than the most profound discourses on such abstract subjects as are beyond the limits of their capacity to understand. (Wakefield 1807: iii-iv)

Wakefield also employs an epistolary framework in *An Introduction to the Natural History and Classification of Insects; in a Series of Familiar Letters* (1816) and *Excursions in North America, Described in Letters from a Gentleman and His Young Companion, to Their Friends in England* (1806).

Harriet Henrietta Beaufort (1778–1865) was born in Ireland, and found an interest in science, especially botany, through the influence and example of her father, Reverend Daniel Beaufort, a founding member of the Royal Irish Academy. An older brother, Francis (1774–1857), became a rear admiral in the Royal Navy and a cartographer and hydrographer, creating the Beaufort scale for wind force, while another, Reverend William (1771–1848) became Rector of Glanmere (Edgeworth 2010). Her sister Frances Anne (Frannie) (1769–1864) was the stepmother of writer Maria Edgeworth. Neither Harriet nor her sister Louisa (1781–1863) married. Their father became increasingly troubled by debt after relinquishing his clerical post in 1818, and Harriet and Louise turned to writing in part in order to support themselves and their father (Harvey and Ogilvie 2000). Both published anonymously, Louisa writing *Dialogues on Entomology* (1819) and Harriet *Dialogues on Botany* (1819). A difficulty with anonymously published books is that they are often ascribed to more well-known authors (as previously noted in the case of Rennie and Granville Penn). Such is the case with *Dialogues on Botany*, which is sometimes attributed to Maria Edgeworth. In her preface, Harriet argues that her mode of adaptation is, in fact, the very use of a dialogue format: "The familiar form of dialogue seems peculiarly suited to the taste of children; they identify themselves with the imaginary characters, and speedily acquire the habit of stating their difficulties and of asking for explanations" (Beaufort 1819: vi).

The work follows the lessons of three siblings, Fanny, Emma, and Cecil, as they are instructed by their aunt, Miss Mary Percival. Active learning is central to this work, for example, in the second chapter where the aunt meets her pupils in the garden (the domestic setting). The children are described as being eager to explore the structure of plants, "Cecil with his microscope, Fanny with pencil and memorandum book, and Emma with her knife well sharpened for the purpose of dissecting" (Beaufort 1819: 7). The author also employs a radical strategy for encouraging her readers to take part in hands-on learning, the omission of illustrations from her book. She defends this by stating that "experience has convinced the author that children should be induced to study nature, rather than engraved representations: every garden, field, and grove, will furnish them with examples; and the habits of examining the structure of real plants, and of seeing with their own eyes, will yield them not only more present pleasure, but far more certain and permanent knowledge" (Beaufort 1819: vi–vii).

A final example is the work of an author who is usually considered to have written in the familiar format genre, but in a style quite different from those discussed so far, Margaret Bryan (fl. 1815). Very little biographical information exists about Mrs. Bryan, so it is not known where her knowledge or independent approach to science writing might have originated. *The Dictionary of National Biography* identifies her as a "beautiful and talented schoolmistress" married to an equally mysterious Mr. Bryan and who taught at several different locations around London (Humphreys 1886: 154). She penned at least three physics and astronomy books for young readers, *A Compendious System of Astronomy in a Course of Familiar Lectures* (1797), *Lectures on Natural Philosophy* (1806), and *An Astronomical and Geographical Class Book for Schools* (1816). An engraving of Bryan and her two daughters appeared in the frontispiece of her first work and pictures her with her left hand resting on the base of a celestial globe, her writing quill in her right hand.[5] Other scientific instruments are included in the picture, including a telescope. The illustration paints her as more than a passive transmitter of knowledge, and indeed Bryan's writings demonstrate that she had considerable skill in the use of astronomical instruments. For example, an 1811 letter to William Herschel discusses her attempts to observe a comet that year (Brück 2009). In her *Lectures on Natural Philosophy* she describes her preparation for writing this book as "eight years' study of the facts I have attempted to investigate, aided by seven years' practical experience to establish these principles" (Bryan: 1806: x).

Bryan's works were well-received by the scientific community, and *A Compendious System of Astronomy* included a letter of support from Charles Hutton, Professor of Mathematics at the Royal Military Academy, who had reviewed the manuscript and rejoiced that "the learned and more difficult Sciences are thus beginning to be successfully cultivated by the extraordinary and elegant talents of the female writers of the present day" (Bryan 1797: xi). The second (1799) edition of *A Compendious System of Astronomy* boasted a list of four hundred subscribers that included the Archbishop of Canterbury and the Secretary at War, as well as Reverend William Lax, Professor of Astronomy at Trinity College, Cambridge, and Reverend Dr. Nevil Maskelyne, the Astronomer Royal.

She explains in her preface to *A Compendious System of Astronomy* that "these Lectures were written for my pupils, and not originally designed for public inspection" (Bryan 1797: viii). This means that any adaptations that were made were meant for the benefit of her students alone, and were not initially meant for public consumption and critique. In particular, Bryan explains to her pupils/readers that the

> study of mathematics would be misapplication of your time, which might be justly attributed to vanity and ostentation, and be considered unbecoming your character as females, by employing that time which is more usefully occupied in pursuits adapted to your situation in society, and as the validity of astronomical computation may be provided by those instruments I have provided, aided by your reason. (Bryan 1797: 103–4)

---

[5] A portrait based on the engraving is held by the National Portrait Gallery, London, and can be found at http://www.npg.org.uk/collections/search/portrait/mw42469/Margaret-Bryan-with-her-daughters?s

But while Bryan might have not taught as much pure math as she might have if her pupils had been male, she apparently had no compunction in teaching applied mathematical techniques to these young ladies (utilizing, as she states above, various "instruments"). *A Compendious System* contains an appendix on the basics of trigonometry, an appendix explaining the mathematical tables found in a popular astronomical ephemeris, and set of trigonometric problems using the quadrant and compass, while *Lectures on Natural Philosophy* concludes with 111 solved astronomical and geographical problems using globes, followed by 20 pages of additional questions. Bryan was clearly setting the bar a bit higher for her own readership, and her books were not only read by a general audience, but were used as textbooks in both girls' and boys' schools.

Bryan's literary style is quite different from most of the works considered so far. At first glance, she might be seen as lecturing to her audience, as noted in the title of her second book. However, her intended audience is clearly her female students, and she speaks to them in first person. For example, she dedicates *A Compendious System of Astronomy* to her students noting "Astronomy being the most important Science I have had the pleasure of introducing to your acquaintance, I think, by publishing my Lectures on that subject, I shall afford you a pleasing retrospect of the sublime ideas it conveys." She refers to herself as "the friend who delivered" these lectures, and notes "to be forgotten by you, would inflict a severe pang on that heart which feels for you almost parental tenderness" (Bryan 1797: iii–iv). In her creation of a conversational tone, the one-sided nature of her dialogue with her pupils is therefore no less befitting of the familiar format label than Harriet Beaufort's epistolary works.

Theological considerations are seamlessly integrated with scientific facts and observations, as is customary in such works. The contemplation of astronomy, in particular, is amenable to such lessons, but it is not only in the celestial that such wonders are to be found. For as Bryan offers,

> whether we soar in contemplation of the majesty and glory of God displayed in the Heavens, or pursue our scrutiny of the wonders and benevolence of his administration and dispensations, manifested in the organization and effects of things upon Earth, still we discover new cause for congratulation, new sources of delight and adoration. (Bryan 1797: 162)

Moral lessons also play a central role in her work, as is common in this genre. *A Compendious System of Astronomy* concludes with the admonition

> I hope with advantage, by planting the infant shoots of that harmonious uniformity, benevolence, and order, you therein observed—and that you will cultivate the impressions you have received, so as to bring forth the fruits of those virtues, let your situation in this sublunary state be what it may, even ever so obscure.... Thus, in all situations, you will find virtue to be your best friend, the most likely to make you happy in yourselves, and loved and respected by the world. (Bryan 1797: 233–4)

Likewise she opens her *Lectures on Natural Philosophy* with an "Address to my pupils," where she explains her purpose for writing this book as to prevent her lessons from fading from her pupils' minds, "arming you with a perpetual talisman, which, your conduct justifying its power, will secure you from all pernicious doctrines, and guard your religious and moral principles against all innovations" (Bryan 1806: iv). Bryan also stresses her own dual roles, those of a good Christian teacher and surrogate mother, to her students:

> Happy indeed am I, in the good effects my precepts have produced on many whom I have had the pleasure of instructing: and on none are these impressions more distinguishable than on my own dear children; in whose strict piety, candid integrity and dutiful affection, I rejoice most exceedingly, and feel more blessed than language can express. That the Almighty may continue to direct my dear children and pupils by his spirit—establish them in the performance of their duty by his aid—and endow them with constancy in bearing the afflictions of human life, is my most ardent supplication! Impressed with this hope and animated by the tenderest affection—I rejoice in the titles of Parent and Preceptress. (Bryan 1806: vi–vii)

Despite her strict adherence to the expected moral and ethical conventions of her day, Margaret Bryan was forward-thinking in a number of ways. Not only was she a woman teaching other women in a field that was decidedly male, but she educated her students for not only a single moment, but in her desire for them to return to her lessons, for a lifetime. In this way Bryan sought to instill an appreciation for lifelong learning in her students, who she considered as dear to her as her own children. Having explored these examples, we now turn our attention to the bulk of the book, a close examination of the lives and works of six women who wrote books in the familiar format in the geological sciences, and situate their work against writings of the same period by women who elected not to use this format for their own work.

References

Aikin J, Barbauld AL (1805) Evenings at home; or, the juvenile budget opened, vol 5, 6th edn. Johnson, London

Aikin J, Barbauld AL (1819) Evenings at home; or the juvenile budget opened, vol 1, 12th edn. Baldwin, Cradock, and Joy, London

Bakewell R (1819) An introduction to mineralogy. Longman, Hurst, Rees et al, London

Beaufort H (1819) Dialogues on botany, for the use of young persons. Hunter, London

Bensaude-Vincent B (2009) A historical perspective on science and its 'others'. Isis 100:359–368

Bratton B (2013) We need to talk about Ted. https://www.theguardian.com/commentisfree/2013/dec/30/we-need-to-talk-about-ted . Accessed 15 Mar 2017

Brück MT (2009) Women in early British and Irish astronomy. Springer, Dordrecht

Bryan M (1797) A compendious system of astronomy, in a course of familiar lectures. Leigh and Sotheby, London

Bryan M (1806) Lectures on natural philosophy. Davison, London

Cooter R, Pumphrey S (1994) Separate spheres and public places: reflections on the history of science popularisation and science in popular culture. Hist Sci 32:237–267

Ellison K (2004) Science and celebrity. Front Ecol Environ 2(9):504

Fyfe A (2003) Science for children. In: Fyfe A (ed) Science for children, vol 1. Thoemmes Continuum, Bristol, pp xi–xxii

Fyfe A (2008) Tracts, classics and brands: science for children in the nineteenth century. In: Briggs J, Butts D, Grenby MO (eds) Popular children's literature in Britain. Ashgate, Hampshire, pp 209–228
Gates BT (1998) Kindred nature: Victorian and Edwardian women embrace the living world. Chicago UP, Chicago
Gavroglu K (2012) Science popularization, hegemonic ideology and commercialized science. J Hist Sci Technol 6:85–99
George S (2011) Epistolary exchange: the familiar letter and the female botanist, 1760–1820. J Lit Sci 4(1):12–29
Godding DW (1847) First lessons in geology. Parsons, Hartford
Harvey J, Ogilvie M (2000) The biographical dictionary of women in science, vol 1. Routledge, New York
Hilgartner S (1990) The dominant view of popularization: conceptual problems, political uses. Soc Stud Sci 20:519–539
Humphreys J (1886) Bryan, Margaret. In: Stephen L (ed) Dictionary of national biography, vol 7. MacMillan, New York
Issitt JR (2002) Jeremiah Joyce: science educationist. Endeavour 26(3):98
Jacson ME (1797) Botanical dialogues, between Hortensia and her four children. Johnson, London
Joyce J (1852) Scientific dialogues: intended for the instruction and entertainment of young people. Knight and Son, London
Kohlstedt SG (1990) Parlors, primers, and public schooling: education for science in nineteenth-century America. Isis 81(3):424–445
Larsen K (2014) Margaret Bryan and Jane Marcet: Making space for 'space' in British women's science writing. In: Narain M, Gevirtz K (eds) Gender and space in British Literature, 1660–1820. Ashgate, Farnham, pp 67–82
Leach C (2006) Religion and rationality: Quaker women and science education 1790–1850. Hist Educ 35(1):69–90
Leane E (2017) A brief history of Stephen Hawking's blockbuster. Nature 541:28–29. doi:10.1038/nature16881
Levy M (2006) The radical education of Evenings at Home. Eighteenth-Century Fiction 19(1):123–150
Lightman B (2007) Historians, popularizers, and the Victorian scene. In: Lightman B (ed) Victorian popularizers of science: designing nature for new audiences. Chicago UP, Chicago, pp 1–38
Lowry D (1822) Conversations on mineralogy. Hunt, Philadelphia
Lyell KM (ed) (1881) Life, letters, and journals of Sir Charles Lyell. Murray, London
Marcet J (1817) Conversations on chemistry, vol 1, 5th edn. Longman, Hurst, Rees et al, London
Murray CM (1808) The British garden, vol 1, 3rd edn. Wilson, London
O'Connor R (2007) Young-earth creationists in early 19th-century Britain? towards a reassessment of 'scriptural geology'. Hist Sci 45:357–403
O'Connor R (2013) Earth on show: fossils and the poetics of science 1802–1856. Chicago UP, Chicago
Percy J (1997) An unrecognized novelist: Frances Jacson (1754–1842). Br Libr J 23(1):81–97
Rennie J (1828) Conversations on geology. Maunder, London
Shteir AB (1990) Botanical dialogues: Maria Jacson and women's popular science writing in England. Eighteenth-Century Studies 23(3):301–317
Shteir AB (1996) Cultivating women, cultivating science. Johns Hopkins UP, Baltimore
Shteir AB (1997) Elegant recreations? configuring science writing for women. In: Lightman B (ed) Victorian science in context. Chicago UP, Chicago, pp 236–255
Telescope T (1803) The Newtonian system of philosophy. Johnson, Philadelphia
Wakefield P (1798) Reflections on the present condition of the female sex; with suggestions for its improvement. Johnson, London
Wakefield P (1799) Mental improvement, or the beauties and wonders of nature and art, 1st edn. Greene and Sons, New Bedford

Wakefield P (1805) Domestic recreation; or dialogues illustrative of natural and scientific subjects. Darton and Harvey, London

Wakefield P (1807) An introduction to botany, in a series of familiar letters, 5th edn. Darton and Harvey, London

Whitley R (1985) Knowledge producers and knowledge acquirers: popularisation as a relation between scientific fields and their publics. In: Shinn T, Whitley R (eds) Expository science: forms and functions of popularisation. Reidel, Dordrecht, pp 3–28

Chapter 3
Jane Haldimand Marcet (1769–1858): The Queen of the Familiar Format

3.1 Biographical Overview

The queen of the familiar format scientific text is generally agreed to be Jane Haldimand Marcet (1769–1858). One of twelve children born to Swiss parents, wealthy London merchant Anthony Francis Haldimand and his wife Jane, she was educated at home by tutors in subjects ranging from astronomy and mathematics to history, Latin, and the arts (Polkinghorn 1993; Rosenfeld 2001). She was also exposed to the social scene in Geneva, where she frequently visited her paternal relatives. At age 15 Jane assumed the role of lady of the house (and surrogate mother to her brothers), including serving as hostess at her father's frequent dinner parties, when her mother passed away due to complications from childbirth (Polkinghorn 1993). Her father was said to have been a major influence on her, "a man of learning and culture, an ardent lover of truth, and hater of all pettiness [who] strengthened and deepened the naturally and serious nature of the young girl. Though not deficient in the sense of humour, she could never throughout life reconcile herself to the least appearance of insincerity, even in jest" (Field 1895: 344).

There is no particular mention of a childhood curiosity in science, but such an interest blossomed after her marriage (at the relatively advanced age of 30) to Alexander John Gaspard Marcet, an Edinburgh-educated Swiss-born physician who conducted research in chemistry and lectured at Guy's Hospital in London and later became a Fellow of the Royal Society. At the time, Alexander Marcet already had a colorful history. Originally training to become a lawyer in Geneva, he and a friend became embroiled in politics and were imprisoned. They were successfully released from prison but received 5 year banishments from Switzerland in return, which is how he found himself receiving a medical degree in Scotland (Coley 1968). He specialized in urinary calcium deposits, and utilized chemical analysis to diagnose these maladies. In particular, he developed a technique using a blowpipe to analyze the chemical make-up of the offending stone (Rosenfeld 2001). His wife undoubtedly benefited from both his interest and expertise in chemistry.

K. Larsen, *The Women Who Popularized Geology in the 19th Century*,
DOI 10.1007/978-3-319-64952-8_3

The Marcets had three children who survived to adulthood, one of whom, François (1803–1883) became a professor of physics in Geneva and fellow of the Royal Society, and in 1842 did groundbreaking experimental work on the superheating of water (Chang 2004). Sophia (1809–1877) married magistrate Edward Romilly (1804–1870) in 1830 and lived in London (James 1996). Louisa (1807–1834) married Eugene de la Rive (1804–1872), brother of physicist Arthur Auguste de la Rive, a colleague of her brother's. They, like François, settled in Geneva (Knowles 2017). The Marcets were active members of the intellectual social circles in both London and Geneva and their close friends included scientists Humphrey Davy (1778–1829) and William Hyde Wollaston and political economist Thomas Malthus (1766–1834), as well as women who were noted authors in their own right, including science writer Mary Somerville, novelist and children's author Maria Edgeworth, and noted social theorist Harriet Martineau (Shteir 1996). The Marcet children were personally vaccinated against smallpox by another member of the social circle, Dr. Edward Jenner (Polkinghorn 1993).[1] Jane also increased her own scientific knowledge through attending lectures at the Royal Institution in London, including Thomas Young's lectures in physics and Humphrey Davy's in chemistry.

She began writing popular-level science books for the education of women and others who attended these lectures and other public science demonstrations, as well as for the education of older children, reportedly with her husband's encouragement. While the work on physics (*Conversations on Natural Philosophy* [1819]) was actually written first, it was *Conversations on Chemistry, In Which The Elements of That Science are Familiarly Explained and Illustrated by Experiments* and *Intended More Especially for the Female Sex* (1806) that was published first and received both critical acclaim and a wide readership. It was reported in 1870 that "160,000 copies of this work had been pub. [sic] in the United States in the year 1853" (Allibone 1870: 1218). While *Conversations on Natural Philosophy* is not as famous as its chemical cousin, it, too, was well-accepted, and was utilized as a textbook, especially in America. Other works include *Conversations on Political Economy* (1816), *Conversations on Evidences of Christianity* (1826), and *Conversations on Vegetable Physiology* (1829).

Each of these *Conversations* was written in dialogue form, and features a female tutor/teacher, named Mrs. B instructing two sisters, Emily and Caroline. The exception is *Conversations on Evidences of Christianity,* in which it is the father who instructs the children. This last work is therefore disqualified from being considered to be truly written in the familiar format. The name *Mrs. B* is nowhere explained or expanded, leading some contemporaries of Marcet to suggest that these works (originally published anonymously) were in fact the works of Margaret Bryan. For example, in his preface to the 8th American Edition of *Conversations on Natural Philosophy* (which he edited without her permission), Rev. John Lauris Blake incorrectly refers to the author of the work as "Mrs. Bryan" even while lauding the author for her pedagogical skills:

[1] Excerpts from Jane and Alexander Marcet's letters and diaries can be found in Polkinghorn (1993).

> Instead of exhausting the intellectual energies of youth in committing to memory definitions and mathematical demonstrations, which would not be understood, she proposed to illustrate the great principles of Natural Philosophy by comparisons of the most familiar kind; and it is believed, Mrs. Bryan has done more, in this way, towards giving youth a taste for the study of philosophy more than all others who have published treatises on the subject. (Blake 1826: iii)

While it is certainly possible that Marcet named her main character in tribute to Bryan (as suggested by Armstrong [1938] and Lindee [1991], among others), the two women's writing styles are very different. Ironically, Blake himself produced several editions of many of Marcet's *Conversations* that had his name rather than hers prominently displayed on the title page (along with adding examination questions, notes, glossaries, or other additions). Such problems with American editions will be examined in Sect. 3.3. Another suggestion for the genesis of Mrs. B is Mrs. Barbauld, co-author of *Evenings at Home* (Shteir 1990).

As has been explored by Brück (2009), the publication of *Conversations on Natural Philosophy* after a long delay is certainly intriguing. In her preface to the delayed work Marcet notes that

> The encouraging reception which the Conversations on Chemistry, and Political Economy, have met with, has induced her to venture on publishing a short course on Natural Philosophy; but not without the greatest apprehensions for its success. Her ignorance of mathematics, and the imperfect knowledge of natural philosophy which that disadvantage necessarily implies, renders her fully sensible of her incompetency to treat the subject in any other way than in the form of a familiar explanation of the first elements, for the use of very young pupils. (Marcet 1824: iii)

Was it simply her own self-consciousness over her lack of mathematical expertise that stayed her hand in offering the work to the public, or was it also the imposing nature of physics and astronomy, and their reputation as the province of men. Certainly Margaret Bryan's earlier works had proven that a woman could successfully publish a general interest book in these subjects, and that such a book did not require the inclusion of mathematics (if one ignores the optional appendices of Bryan's works). If Marcet was initially so hesitant about publishing her introductory work on physics, what event served as the impetus for her to overcome her fears? Brück (2009) suggests that the catalyst for Marcet's change of heart was her first extended meeting with Mary Somerville in Geneva circa 1818, a meeting which was the foundation of an afterwards lifelong friendship. Somerville had already acquired a reputation for being skilled in mathematics and astronomical theory, and her opinion and encouragement would have carried considerable weight. Of that fateful meeting, Somerville wrote in her memoirs,

> At Paris I equipped myself in proper dresses, and we proceeded by Fontainebleau to Geneva, where we found Dr. Marcet, with whom my husband had already been acquainted in London. I, for the first time, met Mrs. Marcet, with whom I have ever lived on terms of affectionate friendship. So many books have now been published for young people, that no one at this time can duly estimate the importance of Mrs. Marcet's scientific works. To them is partly owing that higher intellectual education now beginning to prevail among the better classes in Britain. (Somerville 1874: 113-4)

While Jane's two most well-known works were written fairly early in her marriage, the bulk of her writing was done after Alexander's sudden death from a stomach ailment in London in 1822, after which she split her time between London and Geneva (Rosenfeld 2001). Although her early works had brought her commercial success, she did not have to rely upon them for support during her widowhood, due to a significant inheritance that she had received from her father several years prior. Ironically, it was only after his father-in-law's death that Alexander had retired from medicine in order to devote more time to experimental chemistry, but he would only enjoy a few brief years of retirement (Martineau 1869).

Marcet continued to attend public lectures after her husband's death. For example, by her own admission her two volume *Conversations on Vegetable Physiology* was "almost exclusively taken from the lectures of a distinguished professor of Geneva" (Marcet 1829: v). However, while *Conversations on Evidences of Christianity* (1826), and *Conversations on Vegetable Physiology* (1829) were written after becoming a widow, the overwhelming majority of her later works were aimed at younger children, and represent a wider diversity of literary styles. These include children's stories (e.g. *The Seasons* [4 volumes, 1832–3]) and books on grammar (e.g. *Game of Grammar* [1842]) and history (e.g. *History of England* [1842]). In all cases, Marcet attempts to make the subjects taught particularly attractive and engaging for children. *Bertha's Visit to Her Uncle in England* (1830) is a series of fictional journal entries written by a young English girl who had spent her childhood in Brazil due to her father's military service. Upon his sudden death, Bertha is sent on an extended visit to her mother's brother in England and subsequently sends these entries to her mother in batches, including what she observed and experienced at sea. In her preface to *Mary's Grammar; interspersed with stories, and intended for the use of children* (1835: iii–iv) Marcet offers

> I have so often pitied children who have been studying a grammar which they did not understand, that I thought I could not do them a better service than endeavour to render so dry and abstruse a subject easy and familiar.... The stories have been introduced with the view of amusing children during the prosecution of so dry a study; but they may occasionally be used with advantage as parsing exercises.

The work is written as a conversation between 7 year old Mary and her mother, interspersed with fables, fairy stories, and moral tales. A decade later *Willy's Grammar; interspersed with stories for the use of children* appeared, a very similar parallel work featuring young Willy and his teacher, Mr. Thompson,

> adapted to boys instead of girls, and for that purpose it has been found necessary, not only to change the names from the feminine to the masculine gender, but also frequently to make the illustrations more suitable to the ideas and habits of boys. This alteration was suggested to the author by several very competent judges, with the view of rendering the book more generally useful, by introducing it into boys' schools. (Marcet 1845: i)

In all of these works Marcet displays a sympathy for and cognizance of the needs of her intended audience, which may be one reason for her success. One of these later works for younger children, *Conversations for Children; on Land and Water*

(Fig. 3.1) focused on physical geography, and will be analyzed in detail in the following section. Jane Marcet passed away on June 28, 1858 at the home she shared with daughter Sophia and son-in-law Edward Romilly after suffering for many years from a "nervous malady" (Martineau 1869: 75). Marcet's works outlived her in publication, and her prolific and successful career as a writer of books for

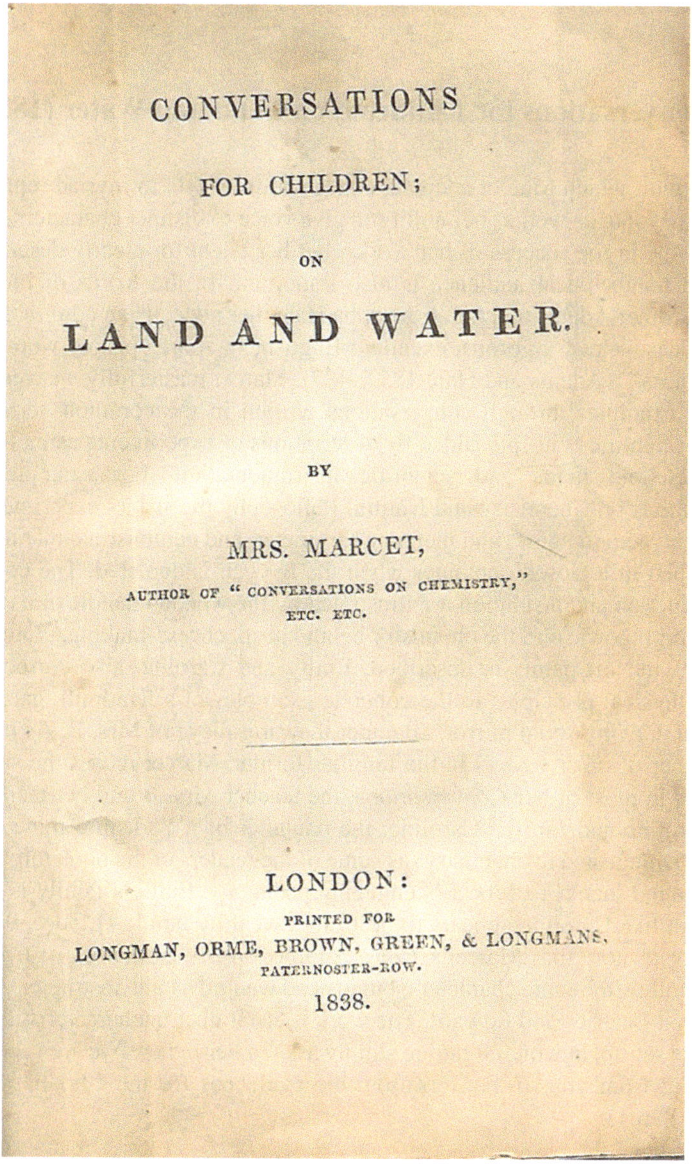

CONVERSATIONS

FOR CHILDREN;

ON

LAND AND WATER.

BY

MRS. MARCET,

AUTHOR OF "CONVERSATIONS ON CHEMISTRY,"

ETC. ETC.

LONDON:

PRINTED FOR

LONGMAN, ORME, BROWN, GREEN, & LONGMANS,

PATERNOSTER-ROW.

1838.

Fig. 3.1 Title page. Reproduced from Jane Marcet (1838) *Conversations for Children; on Land and Water*

young adults and children became, quite literally, the fodder for legends (or at least highly romanticized accounts) by several decades after her death:

> Spending her time partly in London, partly at Geneva, Mrs. Marcet's intense love for young people made her long to share with them the knowledge in which she herself took so much pleasure, and her books, simple, fresh, and easy, while stored with ample and accurate knowledge, exactly met the awakened desire of parents to teach their children the simple facts of everyday life. (Field 1895: 345)

3.2 Conversations for Children; on Land and Water (1838)

The ease with which Marcet adapts a conversational style to myriad topics in science and beyond, as well as her ability to give voice to distinct characters, certainly played a role in the success of her works, but her talent for clearly describing science to a nontechnical audience is also important. In the words of biographers Henry Gardiner Adams and Sarah Josepha Hale, her style "is an admirable vehicle for her ideas—clear, vigorous, excellent English; in short, 'proper words in their proper places'" (Adams and Hale 1857: 487). Marcet masterfully weaves together everyday examples through conversations written in the common vernacular to illustrate scientific concepts, aided by descriptions of experiments using both common household items and scientific instrumentation. For example, in the *Conversations* on Chemistry and Natural Philosophy the sisters experiment with a pyrometer, vacuum pump, and mercury barometer, and demonstrate that a candle is extinguished in a closed container when the oxygen is depleted. The concepts of heat conduction and insulation are illustrated by the wooden handle of a teapot and a flannel nightgown, and the chemistry behind soap, cheese-making, clothing dyes, perfumes, and oil paints is described. Emily and Caroline also correctly apply abstract physical principles to the concrete examples of a windmill, garden water pump and drawing-room mirror, all under the careful eye of Mrs. B. As in the case of a number of other writers in the familiar format, Marcet reuses the same main characters in most of her *Conversations*, the teacher Mrs. B and her teenaged students Caroline and Emily.[2] Caroline, the daughter of a Yorkshire mine owner, is initially disinterested in chemistry (as some of the readers of the book might be), but comes around thanks to Mrs. B's engaging pedagogy. Emily's family is not mentioned, but like Caroline appears to be from a wealthy family (Lindee 1991). It is interesting to note that Marcet's *Conversations on the Evidences of Christianity* does not follow the same characters, but instead we find a male instructor, Mr. B and his children Beatrice and Edward. The work is also rather uncharacteristically awkward in its set-up, in which Beatrice simply asks of her father "We are very desirous of learning from you, in conversation, the evidences for the Christian religion" (Marcet 1826:1).

[2] *Conversations on Political Economy* does not include Emily, but she is mentioned in passing.

As noted in Sect. 2.2, authors using the familiar format often explain early in the work how they intend to deal with issues of vocabulary and any other adaptations. In the case of *Conversations on Chemistry*, Marcet offers quite honestly in her introduction that she presumes her reader to already have some basic knowledge in natural philosophy, and confesses that she herself had prepared such a work, but abandoned her completed manuscript when Joyce's *Scientific Dialogues* had "been pointed out to her, which, on a rapid perusal, she thought very ingenious, and well calculated to answer its intended object" (Marcet 1817: x). This admission certainly speaks to a major reason why that volume had not been published first. She is also cognizant that she sometimes puts questions and explanations into the mouths of her young characters

> which appear much too acute for the young pupils, by whom they are supposed to be made. Of this fault the author is fully aware. But, in order to avoid it, it would have been necessary either to omit a variety of useful illustrations, or to submit to such minute explanations and frequent repetitions, as would have rendered the work tedious, and therefore less suited to its intended purpose. (Marcet 1817: viii–ix)

One common convention of the familiar format is an initial reticence on the part of the young characters, who fear that science is beyond their comprehension. In the first chapter of the *Conversation on Natural Philosophy*, Caroline predictably voices this uncertainty, which allows Marcet (through the voice of Mrs. B) to explain her pedagogical philosophy:

> *Mrs. B.* I make no doubt, my dear, but that I shall be able to explain this difficulty to you; but I believe that it would be almost impossible to render it intelligible to the comprehension of so young a child as your sister, Sophia. You, who are now in your thirteenth year, may, I think, with great propriety, learn not only the cause of this particular fact, but acquire a general knowledge of the laws by which the natural world is governed.
>
> *Emil.* Of all things, it is what I should most like to learn; but I was afraid it was too difficult a study even at my age.
>
> *Mrs. B.* Not when familiarly explained; if you have patience to attend, I will most willingly give you all the information in my power. You may, perhaps, find the subject rather dry at first; but if I succeed in explaining the laws of nature, so as to make you understand then, I am sure that you will derive not only instruction, but great amusement from that study….. (Marcet 1824: 2-3)

As previously explained, central to the mission of science books in the familiar format is the use of science to confirm religious belief and piety; Marcet's works are, quite literally, textbook examples. For example, in *Conversations on Natural Philosophy* Mrs. B. explains to her charges that

> Every step that you advance in the pursuit of natural science, will fill your mind with admiration and gratitude towards its Divine Author. In the study of natural philosophy, we must consider ourselves as reading the book of nature, in which the bountiful goodness and wisdom of God is revealed to all mankind; no study, then, can tend more to purify the heart, and raise it to a religious contemplation of the Divine Perfections. (Marcet 1824: 20)

When faced with the immensity of the universe, Emily proclaims "it seems too sublime for our imagination, and to think that the goodness of Providence extends over millions of worlds throughout a boundless universe—Ah! Mrs. B, it is we only who become trifling and insignificant beings in so magnificent a creation!" Her teacher gently directs her thoughts to what she considers the proper interpretation, that

> This idea should teach us humility, but without producing despondency. The same Almighty hand which guides these countless worlds in their undeviating course conducts with equal perfection the blood as it circulates through the veins of a fly, and opens the eye of the insect to behold His wonders. Notwithstanding this immense scale of creation, we need not fear to be disregarded or forgotten. (Marcet 1824: 160-1)

Mrs. B's lessons in Natural Philosophy predictably end with a reminder to her readers that "in order that the study of nature may be productive of happiness, it must lead to an entire confidence in the wisdom and goodness of its bounteous Author," similar to the admonitions of Margaret Bryan to her students (Marcet 1824: 421).

As noted in the previous section, Marcet turned to writing for younger children after the death of her husband. The most scientific of her later books for young children was *Conversations for Children; on Land and Water* (1838), a work on physical geography that featured two parents (Mr. and Mrs. B) and their four young children (between the ages of six and ten) observing, discussing, and modeling some of the interactions between geological processes and the landscape (such as the erosion of mountains). Marcet was cognizant of her readers' expectations for this volume, given her considerable reputation. She therefore begins her preface with the admonition that

> The Title of this Work may lead my Readers to imagine that it bears some resemblance to my former Conversations; it is therefore necessary to inform them that it is still more humble in its pretensions, and should be considered rather as a continuation of my later works for young children.... I have endeavoured to mingle information with amusement, and to teach those of my young readers who may be learning Geography, that there are other matters connected with land and water quite as interesting as the names and situation of the different parts of the earth. (Marcet 1838: iii)

Marcet demonstrates her comfort and skill with using the familiar format from the first pages, as she sets up the familial setting in the opening sentences:

> Mrs. B had given the first lesson of geography to her two youngest children [Sophy and Caroline], and afterwards sent them to play in the garden with their elder brothers [Edward and William]. They remained there so long that she went out to see what they were doing, and to her surprise found them all busy making mountains.
>
> The gardner had collected in one corner of the melon-ground a store of clay, sand, chalk, and road dust, and with these materials the children were making free, and raising them into heaps of various shapes in imitation of mountains.
>
> "Look, mamma," said Edward, who was beating a large clump of clay with his spade; "this is to be Mont Blanc, which is always covered with snow." (Marcet 1838: 1)

In this initial vignette we see Marcet motivating not only the domestic setting and the female authority figure, but highlighting how natural it is that the children are engaged in hands-on activities and connecting their scientific play to their common experiences. Mrs. B, the scientifically well-versed teacher/mother, inquires as to the activities of her other children, and finds that William is building the Alps, while Caroline is fashioning a mountain that her mother likens to the Apennines in Italy in terms of its shape. When the children take turns using a watering can to make it rain on their respective mountains, Mrs. B points out the different ways the mountains erode (depending on their materials) and notes with interest the way little Sophy's mountain transports water, likening it to a spring.

Over the next few chapters, Mrs. B uses the children's creations as a natural laboratory in which to not only illustrate her lessons on rivers, lakes, islands, and springs, but allow her children to enthusiastically model and experiment. She also utilizes maps in order to compare the mountains of Europe that the children already have some knowledge of with those in North and South America, allowing them to draw connections and apply their newly acquired understanding of geology (Fig. 3.2). Finally, Mrs. B and her children visit a real-life spring as well as a river, as capstone activities to their lessons on these topics. Upon seeing the spring, the children are confused as to where the water (and the pressure to push it up) originate from. Their mother explains that it comes from

> a pond or reservoir of water which is hidden inside the hill you see yonder, about half a mile distant. There was no opening by which the water could get out on the side of the hill, so it went on running down internally till it reached this spot, and here it has made its way out, for you see the hill in which the reservoir is concealed is considerably higher than the ground on which we stand,

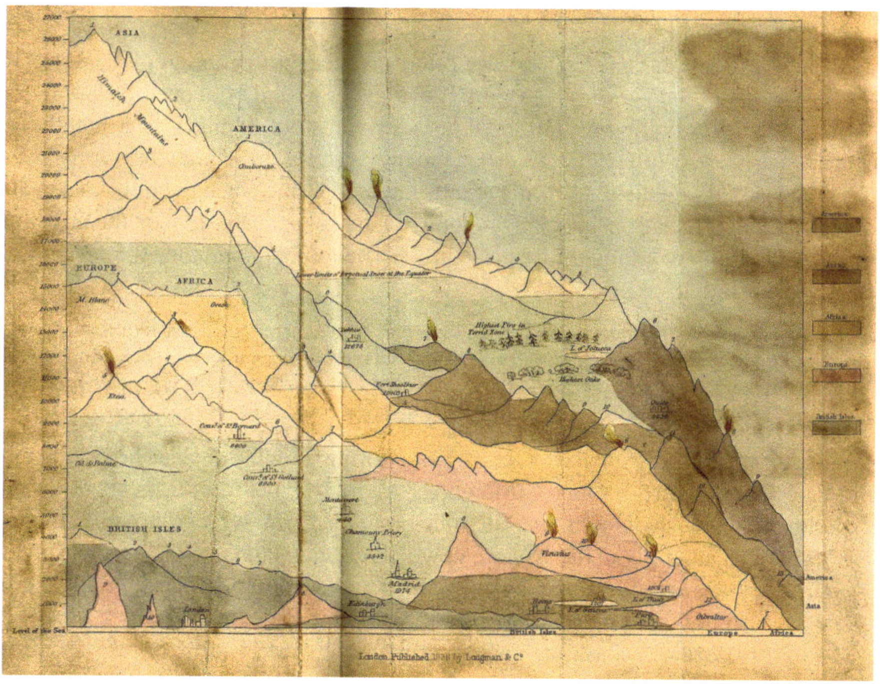

Fig. 3.2 Comparison of different mountain ranges. Reproduced from Jane Marcet (1838) *Conversations for Children; on Land and Water*

but is unable to explain it further "for you are too young to understand the cause of it" (Marcet 1838:70–1). These self-imposed boundaries to her explanation demonstrate a cognizance of the limitations of her target audience and sets the level of adaptation through simplification that she considers proper in this work.

In this related series of lessons (which clearly build upon each other), we see why Marcet was such a popular writer in this genre. The clearly written, flowing dialogue conveys the science in a simplified but scientifically accurate manner. As for the requisite religious aspects of the lessons, they are naturally integrated as well. For example, when William marvels how rivers seem to know the way to the sea, his mother notes that "He who created them… directs their course, by the laws of nature (that is to say, the laws which Almighty God has established to govern nature), water always flows downwards" (Marcet 1838: 34). Similarly, when Sophy expresses surprise that nature takes such good care of simple lichens, her mother counters

> It is not surprising… if you consider that nature means the laws which a perfectly good and Almighty God has established for the preservation of all his creatures; His care is not confined to men and animals, but descends to the lowest amongst created beings, of which the lichen affords a striking example. But small and insignificant as these plants appear, they are, you will find, of great importance in affording the means of forming a soil where none exists; a soil in which, in after ages, the venerable oak may rear its stately head. (Marcet 1838: 222)

Note the connections made between scientific observations, religion, and the everyday mundane world.

Common domestic experiences are often invoked in the children's lessons. A discussion about the pressure caused by lava underground is likened by Sophy to "when the milk is boiling for our breakfasts, it swells out so much, that though the saucepan is not half full at first, if left on the fire, it froths up so high that it runs over the saucepan into the fire." Caroline recalls a specific time when the milk was unattended and after the milk "bubbled and boiled" the cover burst off and "the milk ran over into the fire! We thought it such fun!" Their mother continues the metaphor, explaining

> the lava is very like the milk boiling in the saucepan. When there is no cover it runs over as it does in the crater of a volcano; and when there is no crater open, it makes such efforts to get out that it shakes the ground and sometimes bursts its way through the ground above it; that is, it forces open the cover of the saucepan and overflows. (Marcet 1838: 164–5)

In another example, Sophy wonders "that a spring should be clear after running through so much dirty ground" (Marcet 1838: 144). This leads Mrs. B to do an experiment with a charcoal filter water purifying machine. William notes that there is not charcoal naturally in the ground, to which his mother explains "there are other things which answer the same purpose; gravel and sand, for instance, or any earth that water can trickle through" (Marcet 1838: 145).

Mr. B also makes occasional appearances in the work, but his more masculine demeanor towards both the children and science is clearly evident. In one instance, the children and their mother visit Mr. B in his laboratory, and he takes a few moments from his own investigations to show his children how potassium bursts into flames when placed into water, and how iron fillings and sulphur, when moistened and heated, spontaneously combust, connecting this with lava. The children clamor for more experiments, but their father is clearly impatient, insisting instead "No, my dears, this is chemistry, which you are much too young to understand; and I am not a conjuror to show off a number of wonderful tricks for the amusement of children. So take them away… and let me return to my work" (Marcet 1838: 176). Their father's curtness and seriousness is written in for a purpose, as Marcet undoubtedly had seen similar attitudes in some of the men of science she had known, and perhaps wished to prepare her young readers for such encounters themselves. In depicting such attitudes as rather normal and expected, her readers (both the children and the mothers/tutors/teachers who also read these works) would not take such attitudes as a personal affront or let it dampen their own enthusiasm for science. In a palpable way it also demonstrates the superiority of the mother figure as a teacher, in both her patience and her desire to have the children engage in experimentation to their heart's content.

Marcet also includes examples of the children applying their newfound knowledge to novel situations, for example reasoning through the following hypothetical circumstance:

> "I wonder," said William, "whether there were ever any eruptions of burning lava at the bottom of the sea?"
>
> "If there were it could never be known," observed Edward' "for there would certainly be water enough to put out the fire immediately."
>
> "But suppose, "said Sophy, "there was a very very great eruption from the bottom of the sea, it might throw up ashes and make a Monte Nuovo under the water, which would, perhaps, rise up as high as the surface of the sea, and then it would make a sort of new island…."
>
> "And then," cried Caroline, "if the eruption was strong enough to get the better of the sea, when it was under water, I don't think it would stop when it got above water, and had nothing to hinder it; but it would go on throwing out ashes and lava, and all sorts of things, and so there would be a volcano in the little island, and that would be prettier still!" (Marcet 1838: 188–9)

Edward then comments that his mother has been strangely silent throughout this discussion. "I would not interrupt your discussion," she explains, "because I like to hear you argue without my interference, especially when you are in[sic] the right track" (Marcet 1838: 189). Mrs. B then describes the creation of a temporary volcanic island off the coast of Sicily in 1831. Here Mrs. B. clearly models effective teaching for her audience of teacher-mothers by encouraging the children's application of critical thinking skills to a geological example that is at the same time unfamiliar to them and piques their interest. In the process, the children demonstrate their deep understanding of the topic. Such an example would be appropriate for an introductory level geology or physical geography class even today.

Finally, Marcet is always keenly aware of the entertainment value of her stories, and how children's tales can be used to illustrate scientific principles. A prime example forms the book's final lesson, in which the children wonder about how rocks change as they are worn down, and how new ones made. "I should very much like to know the history of a stone," asks Caroline, "if it could tell whence it came, and whence it is going, and what had happened to it during its travels" (Marcet 1838: 239). Her mother relents, and in a "small voice" tells the first person account of the life (over many, many years) of a stone that started as "part of a large rock, at the top of a high mountain," was broken off by an earthquake and rolled downhill (Marcet 1838). The rock was then buried in the ground, washed by groundwater, dug up by a ploughman and tossed into a pile and finally built into a house wall. When the house eventually fell apart, the freed rock rolled into a stream, where it was eventually dragged downstream in a flood, passing through a delta, and finally made its way into the sea, where it was broken down into sand, ending the tale. Sophy is impressed, but her mother adds that "If I had given you an account of a stone that had found its way into the subterranean fires, and been melted into lava, and thrown up by the crater of a volcano, and formed part of a new mountain, it would have made a much more wonderful tale; but I thought it more useful to you to learn the usual occurrences connected with the existence of a common stone" (Marcet 1838: 248). With this, the tale, and the volume, ends.

As with Marcet's works in general, *Conversations for Children* received generally positive reviews. *The London Quarterly Review* (1844: 12) singled it out as a representative example of "those works which convey more correct information without any expense of interest," calling it "so far superior to the usual class of modern books, in which it is thought necessary to give instruction a garnish of amusement, that, though drawn up in that garrulous form we so much condemn, we cannot omit to recommend it here." The *London Spectator* termed it "A delightfully-entertaining as well as instructive book" (Allibone 1870: 1218). But there was a sinister side to the publishing of popularized works in the nineteenth century, especially in the United States, where authors, even those as famous as Jane Marcet, had no rights over their own works. In the specific case of *Conversations for Children* this led to an unauthorized American edition modified by a male editor to better fit the needs of his audience that was officially sanctioned by a public school board.

3.3 Adaptation and Copyrights in Nineteenth Century America

Harry Gardiner Adams and Sarah Josepha Hale (1857: 487) proclaim of *Conversations on Chemistry* "No work on science in the English language, we might almost say in the world, has been more useful in imparting its knowledge. Its clear elucidation, and its admirably simple method, have undoubtedly contributed, in a great degree, to render chemistry popular." But these biographers also note that

> It is curious to notice the way in which American men have availed themselves of these treasures of intellect without remuneration, or even acknowledgements to the author. Taking these books, and merely giving on the title-page, "By the author of Conversations," &c., they have added "Adapted to the use of Schools," and paraded their own names in full, without an intimation there, or in the preface, that these scientific text-books were the pro-

> ductions of a *lady*! "Give her of the fruit of her hands, and let her own works praise her in the gates," is the command of God respecting woman. In regard to the subject of our sketch, this just tribute has been wholly withheld; yet few scientific writers have so well merited the praise and gratitude of all who read the English language. (Adams and Hale 1857: 487)

For example, as previously explained, Reverend J. L. Blake published an edition of *Conversations on Natural Philosophy* in 1826 under his name (listed as editor) but with the subtle subtitles *By the Author of Conversations on Political Economy* and *Improved by appropriate questions for the examination of scholars; also by illustrative notes, and a dictionary of philosophical terms*. Blake explains in his preface that his supposed improvements were necessitated by his New Hampshire seminary school students' confusion as to exactly what they were supposed to study when using Marcet's book (which, again, was never intended by its author to *be* a textbook).

As curious as it might sound in the twenty-first century, none of this was technically illegal in the nineteenth century. The first American copyright laws were only instituted in 1790. For the next 80 years authors and publishers filed statements of copyright with the offices of district court clerks. In 1870 the Library of Congress took over this responsibility, with the department of the Copyright Office established in 1897 (United States Copyright Office n.d.). It was therefore common to see statements such as the following posted in the first few pages of some of the works discussed later in this book:

> Entered, according to Act of Congress, in the year 1832, by Clapp & Hull, in the Clerk's Office of the District Court of the District of Massachusetts (Welsh 1833: p 4).
>
> Entered according to Act of Congress, in the year 1831, by Mrs. Almira H. Lincoln, in the Clerk's Office of the Northern District Court of New-York (Lincoln 1836: 4).
>
> Entered according to Act of Congress, in the year 1839, by Marsh, Capen, Lyon, and Webb, in the Clerk's Office of the District Court of Massachusetts (Webb 1840: 2).

However, the rights of authors in foreign countries were not protected under this legislation, and foreign authors did not receive any rights until the Copyright Act of 1891, but only in cases where the work was actually printed in the U.S.[3] (Mumford 1952).

This was the environment into which Marcet's *Conversations for Children* was welcomed into the United States in 1838. It was considered such a valuable work for the education of young children that it was adapted for inclusion in the Massachusetts Board of Education's *Juvenile Series* of *The School Library*, in an expanded version of the work crafted without Marcet's input by editor Thomas Hopkins Webb and published under the title *Scenes in Nature, or, Conversations for Children; on Land and Water*, in 1840. In fact, the third of the copyright notices printed above is, in fact, the copyright for *Scenes in Nature*. An analysis of Webb's revision of Marcet's work for an American audience is instructive, as it demonstrates the general lack of respect for intellectual property that was rampant in nineteenth century America (and beyond).

[3] For a detailed examination of the history of copyright law in the United States, see Bracha (2008).

In the late 1830s conventions were held across Massachusetts to discuss the status of public education at which the members of the State Board of Education collected public input. It was determined that there was a need for books, especially for the education of children, and in response, the school districts of Massachusetts were authorized to form school libraries and buy books using a small annual State appropriation. In order to supply suitable books to these libraries, the members of the Board of Education teamed up with Boston publishers Marsh, Capen, Lyon, and Webb to produce two age-appropriate series of 50 books each called *The School Library*. The *Juvenile Series* was aimed at children aged 12 and younger, while the other series was written for older children and their parents (Everett et al. 1840). The two series were a combination of original works by American authors and revised editions of famous works (many of them by European authors), in most cases revised by New England editors.[4]

Each book was officially sanctioned by the State Board of Education before publication, and held to specific standards set forth in the announcement of the creation of *The School Library* in 1840. Because the state was overwhelmingly Christian, but with citizens belonging to a variety of denominations, religious freedom and tolerance (at least as applied to the continuum of Christian beliefs) was highly valued. It was therefore important that no book could be included that could be seen as "an instrument for advancing or depressing the opinions of any sect of Christians" (Everett et al. 1840: 14). The subject of each book also had to be aligned with topics that a "Christian parent would approve" of (Everett et al. 1840: 16). The books were not intended as textbooks, but rather books to be read at a child's leisure, and therefore needed to entertain as well as inform. In a description tailor-made for Marcet's famous works, volumes in *The School Library* were required to

> clothe the subjects discussed, in a popular garb, that they may prove so attractive, as to lure the child onwards, fix his attention, and induce him, subsequently, to seek information from other and more recondite works, which, if put into his hands at the onset, would alarm him, and induce a disgust for that which would appear dry and unintelligible, and of course uninteresting. (Everett et al. 1840: 24–5)

The library markets in the mid 1800s were exceptionally lucrative, which led to the development of a number of similar book series, such as *The School District Library* by New York publishers Harper and Brothers, the *School Library of the American Sunday School Union* (published in Philadelphia), and the *Christian Library*. Therefore the publishers of *The School Library* had a vested interest in making sure that the volumes they selected and produced met the exacting standards of parents, as well as pleasing their intended young audience. In order to accomplish this, a number of volumes were the works of other authors revised by trusted editors in order to make them better fit the tastes of New England families. Four of the first seven volumes in the *Juvenile Series* of *The School Library* fall into this dubious category: three volumes of selections from the well-known works by Jane Taylor, Dr. John Aikin, and Mrs. Barbauld, all edited by Sarah Josepha Hale, and the

[4] For more information on *The School Library* and the specific case of *Scenes in Nature*, see Larsen (2012), from which part of this section is adapted.

seventh, *Scenes in Nature, or Conversations for Children; on Land and Water*. The title page lists no author, but the preface is signed T. H. W., for Dr. Thomas Hopkins Webb, a member of the publishing house.

Thomas Webb (1801–1866) was born in Providence, Rhode Island. After receiving his medical degree from Harvard in 1825 he distinguished himself in literary and historical circles, publishing a number of articles in *Antiquities Americanae* and the *Memoirs of the Royal Society of Northern Antiquaries* (Winthrop 1866). He was also one of the founders of the Providence Athenaeum, the largest library in the state. Upon moving to Massachusetts, he joined the publishing house Marsh, Capen and Lyon, and was elected a member of the Massachusetts Historical Society. In 1850 he was appointed the secretary of the Government Commission for delineating the boundary between the United States and Mexico and shortly afterwards composed the first report on the registration of births, marriages, and deaths for the General Assembly of Rhode Island (Winthrop 1866). Webb was the Secretary of the New England Emigrant Aid Society from 1854 to 1860 and authored the 24 page pamphlet *Information for Kanzas [sic] Immigrants* (1855). The Society aided New Englanders emigrating to the new territory of Kansas, emigration that was encouraged to maximize the number of abolitionists living in the territory in order to discourage slavery from being established there (Johnson 1930). Webb left behind a widow but no children (Quincy 1882).

In his preface to *Scenes in Nature* (1840), Webb explains that the book is

> based upon a small volume, composed by a lady, who has done, and is still doing, much to enlighten and instruct childhood and youth, and indeed many of a still maturer age; - a lady, who, to her credit be it said, has made it the great end and aim of her being, to inculcate correct knowledge and sound principles in the minds of the young; - a lady, who has strong claims upon our good-will and gratitude, for the charms with which she has invested numerous subjects, otherwise dry and uninteresting, and for the happy manner in which she has simplified, and brought down to the capacity of ordinary minds, many matters of science, that, without such aid, would have been far too abstruse for the majority of readers. This lady is Mrs. Marcet.... (Webb 1840: 3)

Webb continues to laud Marcet for several more paragraphs then explains that because her *Conversations for Children; on Land and Water* had been deemed to be

> a valuable, as well as interesting volume for the Juvenile Series of 'The School Library,' we have ventured to add to its pages in order to make it of a suitable size. So numerous and various are the additions, and so interwoven are they with the original text, that we have not deemed it advisable to place the name of Mrs. M on the Title-page; and at the same time, as an act of justice, we would say to the reader, for the information herein contained, you are more largely indented to the Author, than the Editor. (Webb 1840: 4–5)

The result was a volume with nearly 50 additional pages including a glossary and index, and expanded material about North and South America. While Webb's explanation is rife with contrite humbleness, the fact remains that Marcet's work was not only published without her consent, but revised and edited at a level varying from individual word choices to entire passages. At least one of these changes was a significant departure from the intended moral lesson of the work.

The simplest changes are those in spelling, for example where Marcet's labour, coloured, chuse, waggon and vallies become labor, colored, chose, wagon, and valleys. An interesting change in spelling is from Marcet's Himalayan to Webb's Himmaleh, one repeated several times in the volume. Other simple changes in language are also found throughout Webb's revision, including the substitution of paltry for shabby, ingenious for clever, and unpleasant for ugly and nasty. Webb also adds explanatory interjections in places, as if thinking his audience incapable of understanding certain references. For example, Lilliputian is explained as "Lilliputian or small" (Webb 1840: 39) and to references of certain locations in Europe are appended the name of their respective country (such as "the Adige, in Italy" [Webb 1840: 47]). On the other hand, some language is made more formal in Webb's revision. For example, the appearance of lava as being "very much like milk boiling in a saucepan" in Marcet's original (1838: 164–5) becomes "similarly situated to the milk boiling in the saucepan" (Webb 1840: 206). Volcanic bombs are not found "sticking together" in Marcet's work (1838: 172) while in Webb's revision they are not "adhering together" (1840: 214).

In adapting Marcet's work for an American audience, it is understandable that Webb would change examples of landforms found in England to similar structures found in America, especially New England. For example, Marcet's description of vegetables being brought to market in London is simply changed to a similar scene in New York, and a discussion of the high and low tides seen in Aberystwyth is changed to Sciansconsett, Nantucket. The tallest mountain Caroline had ever seen is no longer Snowden, but Round Top in the Catskills, and the mineral springs of Bath and Cheltenham are replaced by Ballston and Saratoga. Webb also adds additional examples, some from Europe, Asia, and South America, but the vast majority from the United States. For example, to the discussion of lakes and rivers he adds New York's Lake George and the Yellowstone River, a tributary of the Mississippi. Boston is included in the section on towns built on hills, and a discussion on the action of the sea on coastlines includes the examples of Cape Cod, Nantucket, and Martha's Vineyard. To the section on landslides (*landslips* as they are termed here) a lengthy (three-page), detailed recounting of an 1826 landslip in New Hampshire is appended.

It is in his lengthy additions that Webb deviates the most from Marcet's original style. While there are a few longer stories within Marcet's work, these are told by Mrs. B as first-person experiences (such as her trip to Mount Vesuvius) or the recounting of a story heard by Mrs. B from an eyewitness (such as a hurricane in the West Indies). These stories always include interjections by the children to break up the narration as well as interesting side explanations. Some of Webb's long narrations are decent copies of Marcet's style (such as his section on the Mississippi), but the majority are rather dry and overly detailed quotations from other authors in the style of straight exposition. An example is a three-page description of the hot springs of Iceland, ironically directly leading into the following original line by Marcet: "The children were now all impatient to hear about volcanoes" (Webb 1840: 196).

Webb uses his role as editor to make corrections to Marcet's original text, as in the case of the discovery of America by Columbus, which he notes was really a rediscovery some 500 years after the arrival of Vikings. Webb also explains that deSaussure was actually the second person to ascend Mont Blanc, not the first. Webb also adds many additional details, such as the exact number of guides deSaussure had with him, the exact chemicals used in making matches, the elevation of Quito and Staubbach in Switzerland, and the color of the sky as seen on Mount Blanc. Both these additions and their lengthy cousins seem in direct opposition to the stated purpose of the series to entertain and leave the child eager to discover further information from "more recondite works" (Everett et al. 1840: 25).

Although Marcet does include religious references in her work, Webb felt the need to add additional examples. When discussing the length of an earthquake, Webb's Mrs. B offers that such disasters "serve as another warning to us… to have our house always in order, and to be at all times prepared for the summons which sooner or later will call us from this world, to the world of spirits" (Webb 1840: 252). Another addition discusses why God would allow such disasters to occur; Webb's Mrs. B quotes an unnamed source as explaining that it is so that "man may feel his dependence upon Him who has all agents under his control" (Webb 1840: 258).

Webb also adds political and social commentary to his edition. In an added section, young Edward explains the origin of the Red Sea's name, and is lauded by his mother for not being one of those typical children who reads a book without retaining its information by the end. An added section on the Mississippi River praises the interdependence of the social classes: "the rich man rolling in his wealth could not, unaided, long survive, any more than the poor man… by an exchange of the money of the one for the labor of the other, benefits result advantageous to both" (Webb 1840: 28). But the most interesting change in the text is an omission from Marcet's original. William wonders why hiring rowboats costs more than using horse-drawn river boats, because "besides feeding the horses and taking care of them, you must buy the horses and you do not buy men." His mother replies "No, thank God! No men are bought in this country; we have no slaves. Men are hired for wages and they work for you or not, as they choose," and then further explains that the wages must be sufficient to cover the worker's basic needs and those of his family. Caroline then interjects that "it would be very wrong not to pay them money enough for all that" (Marcet 1838: 67). Webb's version omits William's reference to buying men, his mother's comment that their country has no slaves, and Caroline's interjection. Instead he goes directly from the comment that horses must be bought to Marcet's explanation of fair wages. These changes are interesting, given Webb's later work for the abolitionist movement in the New England Emigrant Aid Society.

Webb's changes to Marcet's work were a conscious attempt to reshape it according to the Massachusetts Board of Education's opinion of their audience's tastes and sensitivities. Webb's revision was called "very valuable" by *The North American Review* (1841: 260).The lack of an author listed on the title page was also confusing to some reviewers, as in the case of *The Southern and Western Literary Messenger and Review* (1847: 255–6), which referred to the work as a "joint production" between Webb and Marcet. Since Marcet had no say in the production of this volume, it was hardly a joint publication. Sarah Josepha Hale's complicity in *The School Library's* editing of the works of others is surprising, given her protestations less than two decades later concerning the violation of Marcet's intellectual rights. Perhaps it was precisely her role in this series that led to her apparent change of heart.

References

Adams HG, Hale SJ (1857) A cyclopedia of female biography. Groombridge and Sons, London
Allibone SA (1870) A critical dictionary of English literature and British and American authors, vol 1. Trübner and Co., London
Armstrong EV (1938) Jane Marcet and her 'Conversations on chemistry'. J Chem Ed 15:53–55
Blake JL (ed) (1826) Conversations on natural philosophy, 8th Amer. edn. Carter, Boston
Bracha O (2008) The ideology of authorship revisited: authors, markets, and liberal values in early American copyright. Yale Law J 118(2):186–271
Brück MT (2009) Women in early British and Irish astronomy. Springer, Dordrecht
Chang H (2004) Inventing temperature: measurement and scientific progress. Oxford UP, Oxford
Coley NG (1968) Alexander Marcet (1770-1822), physician and animal chemist. Med Hist 12(4):394–402
Everett E et al (1840) The third annual report of the Board of Education. Dutton and Wentworth, Boston
Field EM (1895) The child and his book: some account of the history and progress of children's literature in England. Gardner, Darton, and Co., London
James FAJL (ed) (1996) The correspondence of Michael Faraday, vol 3. The Institution of Engineering and Technology, London
Johnson SA (1930) The genesis of the New England Emigrant Aid Company. N Engl Q 3(1):95–122
Knowles R (2017) What regency women did for us. Pen and Sword, Barnsley
Larsen K (2012) Scenes in nature: science textbooks and the Massachusetts Board of Education's The School Library. Conn J Sci Ed 50(1):17–23
Lincoln AH (1836) Familiar lectures on botany, 5th edn. Huntington, Hartford
Lindee MS (1991) The American career of Jane Marcet's Conversations on chemistry, 1806–1853. Isis 82(1):8–23
Marcet J (1817) Conversations on chemistry, 5th edn. Longman, Hurst, Rees et al., London
Marcet J (1824) Conversations on natural philosophy, 4th edn. Longman, Hurst, Rees et al., London
Marcet J (1826) Conversations on the evidences of Christianity. Longman, Rees, Orme et al., London
Marcet J (1829) Conversations on vegetable physiology, vol 1. Longman, Rees, Orme et al., London
Marcet J (1835) Mary's grammar. Longman, Rees, Orme et al., London
Marcet J (1838) Conversations for children; on land and water. Longman, Orme, Brown et al., London
Marcet J (1845) Willy's grammar. Longman, Brown, Green, and Longman, London
Martineau H (1869) Biographical sketches. Leypoldt and Holt, New York
Mumford LQ (1952) The history of copyright. ALA Bull 46(2):45–47
Polkinghorn B (1993) Jane Marcet: an uncommon woman. Forestwood, Aldermaston
Quincy JP (1882) Memoir of Thomas Hopkins Webb, M.D. Proc of the Mass Hist Soc 19:226–228
Rosenfeld L (2001) The chemical work of Alexander and Jane Marcet. Clin Chem 47(4):784–792
Shteir AB (1990) Botanical dialogues: Maria Jacson and women's popular science writing in England. Eighteenth-Century Stud 23(3):301–317
Shteir AB (1996) Cultivating women, cultivating science. Johns Hopkins UP, Baltimore

Somerville M (1874) Personal recollections, from early life to old age, of Mary Somerville. Murray, London
The London Quarterly Review (1844) Children's books, vol 147. p 12
The North American Review (1841) The School Library 52:257–260
The Southern and Western Literary Messenger and Review (1847) Notices of new works, vol 13. pp 255–256
United States Copyright Office (n.d.) A brief introduction and history. https://www.copyright.gov/circs/circ1a.html. Accessed 29 Mar 2017
Webb TH (ed) (1840) Scenes in nature; or conversations for children; on land and water. Marsh, Capen, Lyon, and Webb, Boston
Welsh JK (1833) Familiar lessons in mineralogy and geology, vol 2. Clapp and Hull, Boston
Winthrop RC (1866) Death of Dr. Thomas H. Webb. Proc Mass Hist Soc 3:274

Chapter 4
Maria Hack (1777–1844): Contrivance and Conversation

4.1 Biographical Overview

The books of Maria Barton Hack (1777–1844) were widely read, but there is limited information concerning her life. She was born at Carlisle on November 10, 1777, to a merchant father Stephen Barton (1755–1789) and his wife Maria/Mary Done (1752–1784), both members of the Society of Friends (Quakers), although her father had converted as a young man and was not, like his first wife, born into the religion (Fell-Smith 1892). Stephen Barton had professed an interest in literature and perhaps even tried his hand at writing, and her paternal grandfather, Bernard Barton, had received a medal from the Royal Society in recognition of a machine he had invented that was used in calico printing. Maria's half-brother, John Barton, was also a writer, producing pamphlets on philanthropic topics including the plight of the poor (Fell-Smith 1892).

Within a short time of the family's 1784 move to London, Maria's younger brother (and later Quaker poet) Bernard Barton (1784–1849) was born, and her mother passed away. Stephen Barton remarried, to Elizabeth Horne (1760–1833), but in 1789 he also passed away, and Maria was taken to Tottenham to live with her stepmother's family. Maria married Chichester merchant Stephen Hack (1775–1823), a fellow Quaker, on November 7, 1800 and bore him four sons and six daughters. Her eldest and youngest sons, John Barton (1805–1884) and Stephen (1816–1894), emigrated to Australia (Danton 1954). Her daughters included Margaret Emily (1814–1886), Elizabeth and Gulielma (*Annual Monitor* 1886).

The first half of the nineteenth century was a turbulent time for British Quakers. Debates over the roles of evangelism and scripture caused schisms within the Society of Friends, not only creating splinter groups, but leading some, like Maria and several of her children, to ultimately join either the Plymouth Bretheren or the Church of England. An important figure in these debates was Manchester minister Isaac Crewdson (1780–1844) and his controversial book *A Beacon to the Society of Friends* (1835). Crewdson argued that the authority of the Scriptures should take

K. Larsen, *The Women Who Popularized Geology in the 19th Century*,
DOI 10.1007/978-3-319-64952-8_4

priority over the nebulous concept of the *Inner Light*, a personal experience with the divine, and pitted a new generation of more evangelical Quakers against more traditional members (Mingins 2003). Within 2 years of the publication of his book, Crewdson, his wife and approximately 300 members had left the Manchester Meeting, and Meetings in other cities, including Tottenham, Bristol, and Birmingham, and attempted to form their own group called the Evangelical Friends, sometimes referred to as Beaconites (Thomas 1912).

Water baptism and communion (referred to as the Lord's Supper) became accepted rites within this movement, and in 1836 or 1837 Maria Hack was baptized by Crewdson and soon after left the Society of Friends. Her Australian sons eventually followed suit, with John joining the Methodist Church and Stephen the Church of England (Danton 1954). The impact of the evangelism movement on Hack can be seen in her *Familiar Illustrations of the Principal Evidences and Design of Christianity* (1824) and *The Christian Ordinances and the Lord's Supper Not Typical Rites* (1837), the latter written shortly after her resignation from the Society of Friends. Her brother Bernard remained a Quaker and felt "puzzled and bewildered among these nondescript novelties; a sprinkling, or watersprinkled, sacrament-taking Quaker is a sort of incongruous medley I can neither classify nor understand" (Barton 1849: 49). Maria Hack spent the last few years of her life in Southampton, dying at Bevis Hill on January 4, 1844.

While Maria was later remembered for her "dignified and lady-like manner" (*Annual Monitor* 1886: 68), it was her prolific pen that garnered her a permanent place in literary history. Her experience in education began at a young age, when she acted as her younger brother's unofficial tutor, and continued into adulthood as she played a role in the education of her own children. She began writing books for children in 1812, including grammar books and primers (*English Grammar* [1812]; *The Little Visitors; a Reader, primary, in words composed mainly of one or to syllables* [1815]) and story books, some of which were history based (e.g. *Grecian Stories* [1819]; *English Stories* [three series: 1820–1825]; *Oriental Fragments* [1828]). She also wrote several geographical works, *Winter Evenings; or Tales of Travellers* (published in four volumes between 1818–24), *The Child's Atlas* (c.1828), and *A Geographical Panorama* (c.1828). She also wrote three science books for children, *Harry Beaufoy; or the Pupil of Nature* (1821), *Geological Sketches, and Glimpses of the Ancient Earth* (1832), and *Lectures at Home* (1834).

Hack often wrote in the familiar format, and like Jane Marcet, reused characters. For example, in *Winter Evenings*, Mrs. B uses conversation to teach geography to her children Harry and Lucy through recounting the adventures of travelers and explorers. Mrs. B becomes the Mrs. Beaufoy in two of the science-based books, where she, and to a lesser extent her husband, teach scientific lessons to their son Harry through hands-on activities and observations. While *Lectures at Home* is not written as a conversation, it is still a work in the familiar format, as the author is having a personal discussion with the reader. For example in a discussion of a piece of glass, she directs her reader to "Take it into your hands, and examine it;- what qualities can you discover by the sense of feeling? It is hard, smooth, and cold. And what can the eye observe? The surface is bright, as well as smooth" (Hack 1834: 5).

Hack found success as an author, and her works generally enjoyed multiple editions (some still in print in revised editions several decades after her death). All of her books for children taught moral lessons and reflected her deep faith, something that was appreciated by her reviewers. For example, an anonymous reviewer in the *Quarterly Journal of Education* lauded at least two of her works, noting

> Her 'Grecian Stories' are full of interest to young children, while her 'Stories from English History' will be perused at a more advanced age with pleasure and profit. These volumes impart correct historical knowledge, and at the same time convey beautiful lessons of morality – the highest and the best use of history. They are written in the form of conversation between a mother and her children; and the machinery is, in this respect, so ably and pleasingly sustained, that the reader has no difficulty in imagining such conversations to have actually occurred. (qtd. in Walker 1843: 8)

In a review of her *Geological Sketches, and Glimpses of the Ancient Earth*, Hack is called "Possessed of a vigourous intellect, which has been actively employed in the acquisition of varied knowledge, and applied to the investigation of truth," and it was observed that she

> might have written for educated adults, had she not modestly preferred instructing children. All her productions, from her Stories of Animals to the volume before us, prove her earnest devotion to the good of the rising generation.... she renders subjects interesting to the juvenile mind, that a common mode of treating would make dry and uninviting. There is, too, in her style a peculiar sweetness of manner, which wins the heart while the information stores the head. (*The Athenaeum* 1832: 77)

At the time of her death, Hack left behind an uncompleted manuscript, another geography text that had been written in the 1830s. Her eighth child, Margaret, had married Thomas Gates Darton (1810–1887), a member of the publishing house Darton and Harvey that had issued some of Hack's books. Margaret completed the book and published it (under her own name but with credit given to her mother) as *The Earth and its Inhabitants* in 1853 (with a revised edition in 1868). In the preface, Margaret explains her mother's reasons for writing the book, providing a succinct summary of Maria's teaching philosophy:

> The great object of this little Work is to make the study of Geography so interesting to Children by lively sketches and entertaining facts relating to the different countries and their inhabitants, that their minds may not only be impressed with geographical knowledge, but awakened to many other subjects of interest associated with it.... When long lists of names are learned by a mere effort of memory, they are often painfully acquired and easily forgotten. Portions only of this Book are intended to be learned by rote; and it is believed that the intelligent Teacher will find these more cheerfully and even better learned by Pupils, whose faculties have been roused and stimulated by description and anecdote. (Darton 1868: iii–iv)

It is a credit to the impact of her children's books that a review of this posthumously published work noted that it was "well written, and interesting, worthy of the daughter of Maria Hack, whose books will always be dear to the young and the old" (*The Gentleman's Magazine* 1853: 70).

4.2 Geological Sketches, and Glimpses of the Ancient Earth (1832)

As previously explained, Maria Hack completed three science-based works. Her first, *Harry Beaufoy*, a natural history volume, will be discussed in the following Sect. 4.3. Her last, *Lessons at Home*, mainly dealt with optics. References will also be made to this volume in the next section. Her middle volume, *Geological Sketches, and Glimpses of the Ancient Earth*, is the only geological text of the three, and will be the focus of the present section.

In her preface to *Geological Sketches, and Glimpses of the Ancient Earth*, Maria Hack seems to presciently contradict the later review of this work that described her works as written for children by explaining "This little volume is not designed exclusively for young persons, but for all to whom the subject is new, and who have not the inclination or opportunity for studying it scientifically" (Hack 1839: iv). She motivates her choice to write in what we now call the familiar format, offering that "the adoption of colloquial intercourse seems to afford the greatest freedom and variety of illustration" (Hack 1839: iv), an interesting contradiction to Lyell's earlier complaint (Sect. 2.2).

As in the case of Jane Marcet, Hack recycles her main character, Harry Beaufoy, in several of her works, a character that not only has a well-developed personality, but grows before the readers' eyes, both intellectually and physically. In her preface to *Geological Sketches* she reminds her readers of the character's salient personal traits and addresses how the years since their last meeting have changed him: "My hero, an intelligent youth of fifteen, having lived with well-informed parents, has acquired habits of thinking and speaking, in some respects, above his years, while he retains the simplicity and eager curiosity of early life…. [Although] he has grown much taller since the last interview" (1839: iv–v). Hack is also careful to describe the parents, both of whom, as seen in Fig. 4.1, take an active role in Harry's education (although in the text clearly more so his mother). It is interesting that while Hack reuses Harry and his parents in two of her scientific works, a sister, Lucy, who was part of *Winter Evenings,* is here omitted and Harry is plainly stated to be an only child. It is explained that Harry had remained at home for an extended period of time rather than being sent off to be educated elsewhere because his father "himself closely engaged in professonal duties, was reluctant to deprive his wife of the pleasure and occupation afforded by the society of her child." For this reason, Mrs. Beaufoy was Harry's primary teacher, his more formal education consisting of merely "two hours in the morning, when he went to receive instruction from the curate of the parish." Hack is clear to explain that this arrangement not only benefitted Harry, but his mother, who, after properly spending the hours when Harry was off with the curate tending to her household responsibilities, afterwards "had ample leisure to read or walk with him, and a powerful motive to improve her own mind, that she might assist in the cultivation of his" (Hack 1839: xvii–xviii). Hack provides

Fig. 4.1 Frontispiece. Reproduced from Maria Hack (1821) *Harry Beaufoy; or, the Pupil of Nature*

a clear argument that rather than diminishing a woman's attention to her customary domestic duties, a mother's teaching of science to her child (and her own increasing of knowledge in the subject through such lessons) instead not only supports the traditional roles of being a good mother and housewife, but augments them (but notice *after* the completion of her domestic chores).

As described by O'Connor (2013), Hack's writing style was more novelistic than Marcet's, with longer descriptive passages and attention paid to the background of both the characters and the setting. Experiments, observations, and even the reading of maps are seamlessly worked into the literary narrative in a natural manner. O'Connor also draws connections between Hack's two Harry Beaufoy-based novels and a geology-based work that is clearly intended for adults rather than children, Charles Kingsley's *Glaucus, or The Wonders of the Shore* (1855). Instead of an eager school-aged child, Kingsley's audience is presumed to be the disaffected tourist on a seaside vacation with his family, to whom the author directly speaks in first person. The tourist/reader is taught to observe the world around him as a means of both finding something of interest in his surroundings, and learning how Darwinism can be reconciled with aspects of religion (Meyers 1989).

As is common with the more fully rendered works of the genre, *Geological Sketches* directly sets up the familiar setting by motivating the first science lesson in response to a normal daily routine. On a typical spring morning, Mrs. Beaufoy asks her son to accompany her on a walk, albeit one longer than usual, to the "top of a neighbouring hill, which commanded a very extensive prospect" (Hack 1839:1). As they enjoy the view and the song of a nightingale, a passing wagon reminds Mrs. Beaufoy of the existence of a nearby chalk pit. Harry is astounded to realize how much chalk had been removed by men over time and its effect on the landscape. His mother characteristically uses the example to direct her son's thoughts to religion, signaling to the reader that, as expected in such works, the geological investigations in this volume will be used to bolster both ethical behavior and faith: "Here is one proof among the many which are continually presented to us, of the goodness of Providence. Chalk, a substance of great importance to man, is placed, you see, in a situation where it may easily be obtained, and in abundant quantity" (Hack 1839: 5).

This initial example also demonstrates the importance of observation and experimentation to the education of young Harry. Their experience near the chalk pit is the impetuous for an entire lesson on chalk, lime and their utility, and Mrs. Beaufoy takes the opportunity to draw another familiar connection, explaining to Harry that she had read in the newspaper that very morning about a fire that was thought to have been caused by a heavy rain reacting with the lime in an outhouse. Harry is skeptical, and his mother offers "if you like we will go to this cottage, and ask the man who has the care of the kiln to give us a piece of quick lime. You may pour some water upon it when we go home, and observe the effect for yourself." Harry is clearly delighted: "Oh! thank you mamma; I like trying easy experiments"(Hack 1839: 11). Unfortunately there are no adults at the cottage when they arrive, only a child who gives Mrs. Beaufoy a piece of chalk instead of quick lime. Harry doesn't understand the difference, until his mother splits open the chalk and finds a fossil shell inside. Harry is now doubly confused as they are many miles from the sea. His mother patiently explains

> I can only explain it by introducing you to a fresh scene of wonders. This shell, my dear boy, is one of a multitude of similar productions: they have been called, and, I think, with great beauty of expression, The Medals of Creation. A medal, you know, is an ancient coin, or a piece stamped to perpetuate the remembrance of some remarkable event. I will give this shell to you, as a medal or memorial, for such it is, of the time when the hill upon which we are now sitting was in a fluid state. (Hack 1839: 18–19)

The revelation that their hill was once part of the seafloor astounds Harry, and sets up further lessons in geology.

In her first chapter Hack has not only introduced her reader to the main characters, their relationship, and the setting in which the main scientific investigations will take place, but basic scientific concepts, the relationship of geology to technology, economics, and, importantly, religious faith, and the process of scientific investigation (including careful observation of the natural world and the role of experimentation). She has accomplished all of this while simultaneously piquing the reader's interest in the topic through the use of Harry's discovery of what appears to him (in his uneducated state, similar to that of the reader) to be mysteries and paradoxes. As a seasoned and previously successful writer of children's works, Hack's literary prowess should not be surprising. However, equally important is the scientific background she acquired prior to publishing this work. Although there is no evidence that she had access to any public lectures on the topic, she was undoubtedly well-read in the subject. A close reading of *Geological Sketches* by Camilla Leach (2006) suggests that Hack consulted nearly 20 contemporary works on both geology and geography, as well as related scientific topics, in preparing this volume.

While Hack had prepared herself for writing this work by reading the standard texts of the day, as an experienced children's author she was also cognizant of her role in digesting and repackaging the scientific content for her audience, something she directly addresses for her audience in the preface. She opens by explaining both her own interest in the topic as well as her philosophy in adapting and simplifying it for the benefit of her reader:

> A traveler, passing through a picturesque country, may add to his enjoyment by delineating the cottage, the rock, or the tree, which happens to please his fancy, without venturing to pourtray [sic] an entire landscape. And thus, while reading the works of eminent geological writers, amused by ingenious theories, and interested by their application to well-authenticated facts, I have sought to perpetuate my own pleasure, and also to contribute something towards the entertainment of others, by attempting a series of sketches, which, without pretending to the completeness of a system, may afford some glimpses of the structure of the Earth. (1839: iii)

Hack returns to her personal philosophy concerning the need at this level for simplification when Harry wants to read some books on the subject to supplement his lessons. Instead, his mother dissuades him, citing the scientific vocabulary that is foreign to him, as well as the fact that "Many young persons call books '*dry*,' that would entertain them exceedingly if they were prepared to understand them. Each art or science has a kind of language belonging to itself, consisting either of appropriate words, or of common words, used in a peculiar sense" (Hack 1839: 55). However, with simplification comes the danger of misconceptions, which Hack is also careful to correct. For example, Mrs. Beaufoy clearly addresses one of Harry's concerning igneous rocks, forcefully stating "I must request you not to use the word *fire* when you refer to Dr. Hutton's theory: *heat* is a more correct expression" (Hack 1839: 114).

Her reference to "Dr. Hutton's theory" (also termed by her *The Huttonian Theory*) refers to her clear and succinct discussion of the history of the Neptunist/Plutonist debate and the syncretism of Hutton. After a brief definition of the two original points of view, Harry asks his mother "Can you tell me whether the Neptunists or the Plutonists are in the right?" His mother offers that she has

> not the presumption to decide that point, Harry. Each party is earnest in defense of its own system, while, as is often the case with other disputes, perhaps the truth may lie between them. Some persons maintain that fire and water have *both* been concerned in producing the present arrangement, and they bring very strong arguments in support of this opinion. (Hack 1839: 60)

A particularly interesting exchange occurs during a discussion of fossils. Mrs. Beaufoy describes how Mary Anning found both Ichthyosaur and Plesiosaur fossils, to which Harry responds "How curious, that she should find both this monster and the fish-lizard!" His mother is the one to broach the subject of gender and geological field work in her reply: "It is certainly uncommon to hear of a lady engaging in such a fatiguing, hazardous pursuit; and I think few would be found willing to undertake a personal examination of the cliffs, especially in the depths of winter," referencing the fact that Anning's major discoveries took place in a January and February (Hack 1839: 308). In comparison, Hack's several pages-long summary of Maria Graham's aforementioned (Sect. 1.4) first person account of the 1822 earthquake in Chile (including properly cited quotations) makes no commentary on her gender, except to note that she was an "English lady then residing in Chili [sic]" (Hack 1839: 125).

The gendering of both science and pedagogy are addressed in other examples within this work. As in the case of Mr. and Mrs. B in Jane Marcet's *Conversations for Children*, there are differences between Mr. and Mrs. Beaufoy's teaching styles (and Harry's response to these differences), which Hack clearly addresses with the apparent intent to prepare her young readers to enter into useful scientific lessons with all types of instructors. In particular, Harry is more guarded with father, and more anxious and hesitant of his own abilities, something his father notices. In response Mr. Beaufoy encourages his son to not hold back, but instead challenge what his father says. In a reference back to one of the central strengths of the familiar format itself, Mr. Beaufoy explains "I wish to converse with you, my boy, not to harangue like a lecturer" (Hack 1839: 326). Hack shares with the audience that Harry (presumably like some of her young readers) had been embarrassed by his "anxiety to please his father, and to sustain his own part in the conversation, without making what he called 'foolish blunders'" (Hack 1839: 327). As Hack explains the reasons for Harry's hesitation to be open with this father, one can imagine the reader nodding in sympathy:

> Mr. Beaufoy, though kind and considerate, had not "a mother's patience" – he had, moreover, a quick sense of the ridiculous, which occasionally showed itself in a smile or tone of irony. In early childhood, Harry's sensitive temperament had often shrunk from this: he was now of firmer nerve, and sincerely loved and honoured his father; but he had not yet learned to feel quite so much at ease with him, as with "that softer friend," who was his daily companion. (Hack 1839: 327)

Harry is eventually encouraged by his father's earnest kindness, and finally overcoming his reticence, the geology lesson continues.

As is usual in a work in the familiar format, the geology lessons often utilize everyday household examples to illustrate major points. For example, Mr. Beaufoy explains of a type of stone found in Purbeck, "though a very hard paving stone, is said to have the property of exuding moisture with change of weather" (Hack 1839: 378). Mrs. Beaufoy, said to have "been attentively listening to the conversation," now offers an important observation of her own: "I suppose that may explain an appearance in the floor of our kitchen, which has puzzled me. I have several times observed some of the stones looking quite damp, as if newly washed, when I knew they had not been wetted. I was not aware of this being a peculiarity in the stone" (Hack 1839: 378–9). Harry is delighted to discover that his mother is "sometimes puzzled by common things! How glad I am that you and papa have taught me to observe and think about them; it has given me so much pleasure" (Hack 1839: 379). In this particular conversation Hack illustrates two very important lessons: that learning is a lifelong process, and the application of the skills of observation and reasoning to everyday life.

Something customary in works of this era was the initiation of chapters with quotations from other authors (often religious in nature). However, in this work these quotations are surprisingly secular, given the overall religious undertone of the work. The intention to utilize geology in the service of religious lessons is established early in the volume. In an early instance, Mrs. Beaufoy exclaims

> We are surrounded by many [wonders] which are beyond our comprehension, but a vast number of interesting facts are open to our inquiries, and as we become acquainted with them, and reflect upon them, we shall gradually form clearer notions of the world we inhabit, and of that series of operations by which the Great Architect has made it what you see. Can you imagine any inquiry more deserving our attention? (Hack 1839: 20)

The book symmetrically ends with a religious and moral lesson from Mr. Beaufoy, who admonishes his son that

> A large fortune does not make a man rich, unless it is managed with economy. The ablest teacher cannot make him wise, if he suffers the streams of knowledge to glide past him unemployed.... Neither the best example nor the soundest instruction — neither enlarged views of Nature nor just conceptions of Providence, will make us *good*, unless these convictions of the understanding are accompanied by the practical submission of our own wills and affections to the influence of religious principles. (Hack 1839: 383)

While religiosity is clearly apparent in this work, Hack does not support a literalist interpretation of the Bible, and is keen to explain to her readers that there is no contradiction between faith and science. In a discussion of the formation of mountains Harry asks his mother if these events happened in the Deluge. Mrs. Beaufoy answers

> It does not seem improbable; but we have no positive information of the changes produced in the structure of the Earth by that awful event.... The short account contained in the early chapters of Genesis seems designed to teach the important truth, that God is the Creator, Governor, and Judge of the Earth and its inhabitants – that His will is the rule of their duty, and that willful disobedience to it will be awfully punished. On subjects not connected with this purpose the historian is silent; and therefore, with regard to the means, the succession of causes, by which the Divine Wisdom has brought this beautiful world to the state in which we see it, man has been left to discover them gradually by the exercise of his reason. (Hack 1839: 32-3)

A footnote connects this discussion to Cuvier, but as explained in Sect. 1.2, religious interpretations of Cuvier's work are largely colored by Jameson's commentary to his translation. Equally important is that Hack does not claim that geology provides definitive evidence of the Flood, only that the two are consistent, and is clear to explain that Scripture is not meant to be read as a complete scientific description of the entire history of the planet. Similarly, when Harry questions early in their lessons "how the gradual formation of land by depositions from water, and the tedious labours of coral-worms, is to be reconciled with the account given by Moses. *Six days!* Why six thousand days would not have done it," his mother summarizes the work of other geologists, such as Buckland, in interpreting the six days of creation figuratively rather than literally and offers that none of these alternate interpretations "has any tendency to lessen your reverence for the Scriptures" (Hack 1839: 38–40).

Nevertheless, the importance of evangelism in Hack's religious philosophy (as well as more general societal prejudices of her time) is apparent in a discussion of coral islands, a topic that Hack discusses in detail "partly because the subject is so very curious, but chiefly because, from the commencement to the perfection of their structure, we are permitted to trace the operation of the means employed by the great Creator for the completion of his work" (Hack 1839: 147). She laments that

> It is indeed melancholy to reflect how long these beautiful islands have been tenanted by a race of people who realized the description of good Bishop Heber: Here 'every prospect pleases, And only man is vile.' But a brighter day is now dawning upon them: the natives are relinquishing their depraving superstitions, and we may hope, that to all the blessings and bounties of Providence, will be added the increasing knowledge of Him from whom they flow. (Hack 1839: 147-8)

Hack specifically addresses the utility of geology in shaping the morality of young persons in an appendix, in a letter written by R. J. Bryce to the Principal of the Belfast Academy. Bryce notes that his school had received donations of samples, "the most remarkable of which was a collection of native shells, containing about a hundred distinct species, gathered by a lady with her own hands, on the beach of Lough Foyle, for the express purpose of being presented to the Museum" (Hack 1839: 389). The identity of this unnamed fossil collector is not specified. Bryce further offers that positive behavioral changes had been noted by the parents of geology students, explaining that it has made the boys "more animated and intelligent; and, by giving them a rational pursuit in their hours of amusement, has done more to 'keep them out of mischief,' than any other thing that could have been devised: for all of them follow up the study more or less, and some so far as to form little cabinets of their own at home" (Hack 1839: 389).

Reviewers of *Geological Sketches* specifically lauded the ethical and religious lessons in her work. *The Gentleman's Magazine* offered that "although the work might probably have been arranged with more judgment, in order to suit the capacity of children, yet the moral tone which pervades the volume compensates for any other defect, and enables us conscientiously to recommend it as a valuable present to the rising generation" (1832: 47). *The Athenaeum* noted that "the adoption of the interlocutory style, affords her an opportunity, which she seldom neglects, of enlivening the technicalities of systems by anecdotes and illustration," but what is "yet more important, she connects mental effort with moral and religious use, and constantly directs the youthful mind, to consider second causes but as effects of the great primary cause; she ever sets forth the great Creator, who, 'in wisdom hath made all his works" (1832: 77). However, a careful consideration of Hack's interpretation of the relationship between science and religion (as well as that of other women who popularized geology) must take into account a particular theological tradition, the Natural Theology of William Paley.

4.3 William Paley and the Education of Harry Beaufoy

All of the women whose works have been discussed thus far have noted the religious lessons inherent in science. Indeed, for some, like Margaret Bryan, the teaching of proper Christian values seemed nearly as important as the learning of science. But there is a particular type brand of religious thought that was highly influential on many women writers in the familiar format, Natural Theology. While the term more broadly refers to the concept of discerning God's wisdom through his plan—the well-ordered world in which we live—the term is most often applied to the book of the same name written by William Paley in 1802. This point is made abundantly clear by the number of times Paley's name and famous examples appear in books in the familiar format. For example, in the preface to her *Lectures on Natural Philosophy*, Margaret Bryan explains that

> Through the whole of the anatomical parts of these Lectures I have followed the very excellent divine Dr. Paley, in his *Natural Theology*: - a work comprehensive in its nature, important in its application, and extensive in its elucidations of the divine wisdom and omnipotence of our great Creator; tending to the subversion of all irreligious principles and doctrines, by the most substantial evidences of the divine origin of religion; and inviting us to that consideration on our nature and our dependence on Divine power and goodness, which cannot fail to confirm us in the true belief and service of God. (Bryan 1806: xi–xii)

In his highly influential *Natural Theology*, Paley takes his reader through a series of arguments intended to demonstrate the irrationality of atheism, and simultaneously convince the reader that the only logic explanation for what he argues is the irrefutable hand of a designer in the world, is the existence of God. He begins with a now famous analogy:

> In crossing a heath, suppose I pitched my foot against a *stone*, and were asked how the stone came to be there; I might possibly answer, that, for any thing I knew to the contrary, it had

> lain there for ever.... But suppose I had found a *watch* upon the ground, and it should be inquired how the watch happened to be in that place; I should hardly think of the answer which I had before given, that for any thing I knew, the watch might have always been there. Yet why should not this answer serve for the watch as well as for the stone? (Paley 1821: 9)

The logical answer is that the watch obviously had a maker, some intelligence that designed and constructed the timepiece for its intended use. Paley then makes the following argument:

> There cannot be design without a designer; contrivance without a contriver; order, without choice; arrangement, without any thing capable of arranging; subserviency [sic] and relation to a purpose, without that which could intend a purpose; means suitable to an end, and executing their office, in accomplishing that end, without the end ever having been contemplated, or the means accommodated to it. Arrangement, disposition of parts, subserviency of means to an end, relation of instruments to a use, imply the presence of intelligence and mind. (Paley 1821:17)

Paley finally argues that the same must be said of the rock; since the natural world shows design, it must also have a designer, but since the design of the natural world is more awe inspiring, so, too, must the power and ingenuity of its maker far exceed that of mere humans.

As one of his prime examples, Paley compares the telescope to the eye, and argues that just as the telescope is clearly designed for the purpose for which it is used, so, too, must the eye not only have been designed in general, but the eyes of different creatures were perfectly designed in order to function under the conditions in which it lives. In Paley's words,

> the rays of light, in passing from water into the eye, should be refracted by a more convex surface than when it passes out of air into the eye. Accordingly we find that the eye of a fish, in that part of it called the crystalline lens, is much rounder than the eye of terrestrial animals. What plainer manifestation of design can there be than this difference? (Paley 1821: 23)

Paley's influence can plainly be seen in this passage from Maria Hack's *Lectures at Home*:

> Compare the globular lens of the fish with the section of the crystalline lens in the human eye. What a difference! and why should this be? It is because the eye of a fish is intended to see in the water, the eye of man to see in air..... A small globe is the most powerfully converging of all lenses; hence it is the most suitable form for the lens of a fish's eye, which has need of great power to collect light in the water. (Hack 1834: 206)

For this reason, many of the familiar format works of the late eighteenth and early nineteenth centuries fall under the secondary genre of *narratives of natural theology* (Gates 1998).

Perhaps the most popular and successful work of this genre was Maria Hack's *Harry Beaufoy; or, the Pupil of Nature*. As a review in Reverend Samuel Wood's *Prayers for the Use of Sunday Schools with Lists of Books Suitable for the Children* opined "This is one of the most interesting works that ever was written for children. It is an excellent introduction to Dr. Paley's Natural Theology (1829: 58). Hack's debt to Paley is evident from the title page (Fig. 4.2), which carries the following

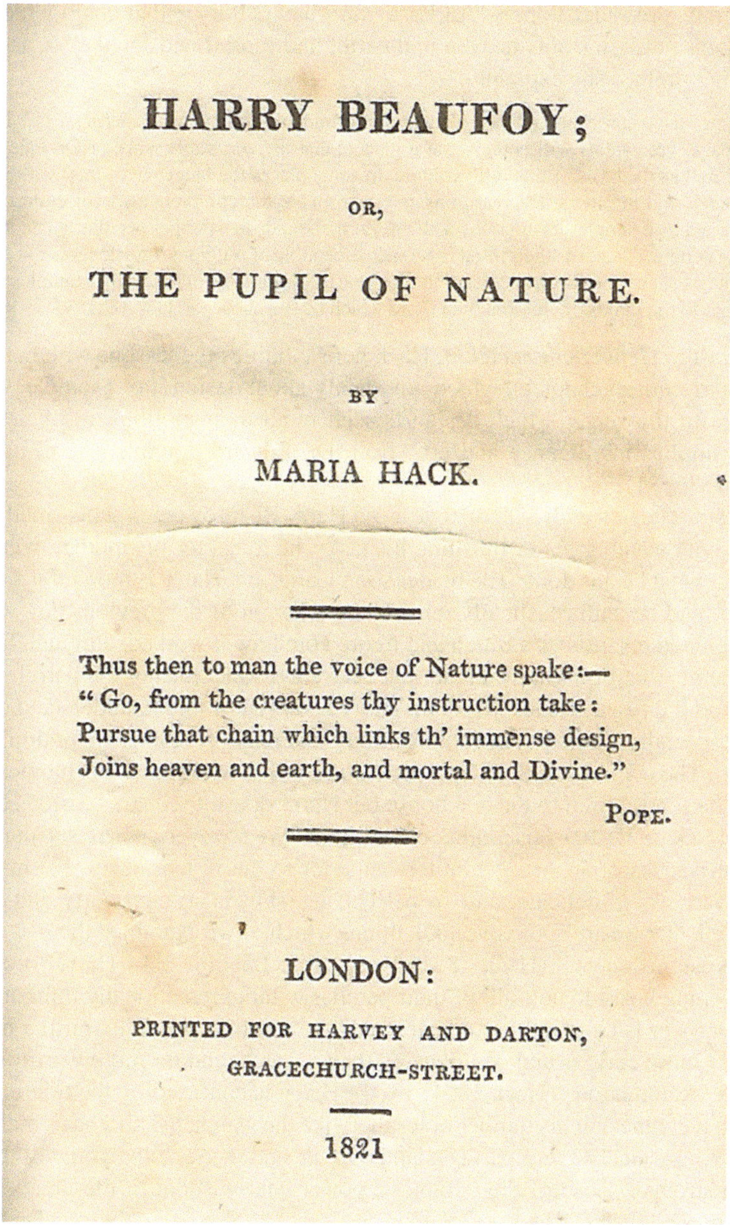
HARRY BEAUFOY;

OR,

THE PUPIL OF NATURE.

BY

MARIA HACK.

Thus then to man the voice of Nature spake:—
"Go, from the creatures thy instruction take:
Pursue that chain which links th' immense design,
Joins heaven and earth, and mortal and Divine."

POPE.

LONDON:

PRINTED FOR HARVEY AND DARTON,
GRACECHURCH-STREET.

1821

Fig. 4.2 Title page. Reproduced from Maria Hack (1821) *Harry Beaufoy; or, the Pupil of Nature*

quote from Alexander Pope's "An Essay on Man": "Pursue that chain which links th' immense design,/Joins heaven and earth, and mortal and Divine" (Cary 1841: 85). In her preface she explains

> Though I have purposely avoided placing the formidable words "*Natural Theology*," in the title-page, yet parents will perceive, at a glance, that the admirable work of Dr. Paley has been used as the basis of this little volume. In some places the language of that work is too technical; and in others, the reasoning is either above the comprehension of children, or illustrated by examples unsuitable for them, yet it abounds in passages peculiarly adapted to young people, as soon as their minds become susceptible of religious knowledge.... In doing this, it has sometimes been necessary to illustrate the arguments of Paley, by examples more familiar to the minds of children than those which he has chosen. (Hack 1821: v–vi)

In contrast to *Geological Sketches*, Hack here solidifies the lessons with her quotes at the start of each chapters, which are mainly about design. For example, Chap. 1 begins with the verse "Much design/Is seen in all their motions, all their make:/ Design implies intelligence and art," from Dr. Edward Young's *The Consolation* (Grigg 1836: 81).

The story begins with a then 10 year old Harry Beaufoy tending his small garden one summer evening. After finishing his tasks, he looks for his mother, wanting to show her what he has done. The reader soon learns that Harry is more of a doer than a reader, and he enthusiastically remembers a lesson of the previous day when he had examined his mother's watch and figured out how it worked, although he "cannot remember half of what we read; but I have a distinct idea of the barrel, and the chain, and the fusee" (Hack 1821: 6). Harry adds "But I like to understand contrivances, especially when they are so ingenious, and really set things in motion" (Hack 1821: 8). The connection to Paley's watchmaker God is obvious. His mother has an idea—she sends him into the hen house but Harry doesn't see any "curious contrivance" (Hack 1821: 10). His mother chides him: "We are every where surrounded by the contrivances of an Artist infinitely superior to the watch-maker; but his works are so various, so delicate, and so perfect; his skill is so constantly and silently employed, that we are apt to overlook things which, if we examined them, would fill us with astonishment." (Hack 1821: 10-1). Mrs. Beaufoy asks Harry to consider why sleeping birds do not fall off their perches without intentionally thinking about it. The answer is that God has constructed "the machine we call a bird: a machine infinitely more curious and complicated than a watch, and far more worthy of your attentive examination" (Hack 1821: 13–4). Harry is confused by the reference of a bird as a machine, but his mother notes that, like the watch, the bird has a maker, and "observe how nicely every part is adapted to the end in view, and you will then feel a much stronger sense of the goodness, power, and wisdom of the great Artificer who contrived them" (Hack 1821: 15).

Mrs. Beaufoy then asks the cook, Betty, to bring in a dead chicken and does an experiment for Harry. When the legs are bent, the claws close as if grasping the perch. Mrs. Beaufoy argues that the tendons must have been designed for this purpose, and "when we plainly see a contrivance adapted to any particular purpose, we are *quite sure* that some intelligent person has been employed in designing and perfecting that contrivance" (Hack 1821: 20). The next day Harry and his mother dissect the leg of the bird, to examine the tendon described the previous night.

Notice not only the mother-led observation and experimentation, but the continual reference to contrivance (one of the central terms used by Paley himself) and the Designer Deity. Mrs. Beaufoy later uses the anatomy of the human backbone to demonstrate that people are contrivances as well, carefully designed by their creator to be able to move in particular ways. For example, she explains how God made humans unable to turn their heads 180° so they cannot harm themselves, evidence of the benevolence and wisdom of God. Similarly, Mr. Beaufoy describes how human bodies are designed to avoid the dislocation of a shoulder under normal circumstances (this particular discussion brought on by a newspaper article describing a stagecoach accident). Other physiological examples of benevolent design include how arteries are "deeply seated; while the veins, which may be wounded with much less danger, lie above the arteries, come nearer to the surface, and are more exposed, Here is a proof of care and contrivance, that cannot be disputed or denied" (Hack 1821: 106).

But recall that for Paley it is not merely the biological that bears witness to the Creator, but the geological world as well. William Buckland had applied Paley's arguments to the mineral world in a famous work published only a year before *Harry Beaufoy*, the aforementioned *Vindiciae Geologicae* (1820). Buckland sees in the geographical distribution of minerals, including coal, salt, and metallic ores, evidence of "design and benevolent contrivance, [and] we trace the finger of an Omnipotent Architect providing for the daily wants of its rational inhabitants, not only at the moment which he laid the first foundations of the earth, but also through the long series of shocks and destructive convulsions which he has caused subsequently to pass over it" (Buckland 1820: 12). Buckland argues that "Geology contributes proofs to Natural Theology strictly in harmony with those derived from other branches of natural history" which, although "less numerous and obvious… they are calculated to lead us a step farther in our inferences" (Buckland 1820: 18).

It is therefore not surprising that Maria Hack's *Geological Sketches* includes a quote from Paley's *Natural Theology* (1821:40) on its title page: "Whatever is done, God could have done without the intervention of instruments or means; but it is in the construction of instruments, in the choice and adaptation of means, that a creative intelligence is seen." The previously described recap of Harry's personality includes the explanation that in his parental lessons he

> was encouraged to exercise the dawning powers of reason in observations tending to show that the marks of *design* and *contrivance* exhibited in the structure and habits of animals, are of such a nature as to assure us that benevolence, or the purpose of communicating happiness, must have been the motive which influenced the great Creator to bestow the gift of life on the countless multitude of beings he has formed. (Hack 1839: xv; emphasis original)

As in the case of Paley's work, and Hack's later *Lectures at Home*, eyes are included as an important example of design, in this case the eye socket of a fossil ichthyosaurus, which consists of "a circle of bony pieces, surrounding the space where the eye has been," termed a common feature of lizards, tortoises and birds but not fish or crocodilians (Hack 1839: 296). Harry considers it strange that this "fish-lizard is in other respects so like a crocodile, and so *un*like a bird! This does not seem like the uniformity you often speak of as prevailing in the works of Providence." But Mrs. Beaufoy is quick to address her son's doubts, offering that "It may appear so at first sight; but you may be sure, whenever this is the case, that the defect lies, not in the contrivance of the Artist, but in our comprehension of his design" (Hack 1839: 297). She asks Harry to consider why this bony rim exists in some cases but not others, and explains that the fish-lizard was designed to swim through the water very fast like a bird flies through the air, thus it needs vision like a bird.

Mrs. Beaufoy also invokes Paley when discussing past climates, musing that Paley might have explained that

> In this distinct impression of a tropical fern, I not only see a strong probability of a high and uniform temperature having once prevailed over the Earth, but I see clear evidence that, in a former state of things, so different from that now existing, we can but faintly imagine it, the same Great Architect was silently at work, preparing, by means similar to those which he now employs, for a future and grander development of his power and goodness. In the delicate veins of this fern I acknowledge the same Hand that arranged the fibres of the living plant.... (Hack 1839: 221)

Hack's specific use of the arguments of William Paley within her works in the familiar format can be seen echoed in the books of other women writers in this genre, in their attempt to demonstrate that geology is the willing handmaiden to Natural Theology. In doing so, these women justified not only their very writing of these works, but the time and effort of their impressionable young readers.

References

Barton L (ed) (1849) Selections from the poems and letters of Bernard Barton. Hall, Virtue and Co., London

Bryan M (1806) Lectures on natural philosophy. Davison, London

Buckland W (1820) Vindiciae geologicae. Oxford UP, Oxford

Cary HF (ed) (1841) The poetical works of Alexander Pope. Smith, London

Danton L (1954) The baptism of Maria Hack, 1837. J Friends' Hist Soc 46(2):67–77

Darton ME (1868) The earth and its inhabitants. Routledge, London

Fell-Smith C (1892) Maria Hack. The Athenaeum 24:889

Gates BT (1998) Kindred nature: Victorian and Edwardian women embrace the living world. Chicago UP, Chicago

Grigg J (1836) The poetical works of Milton, Young, Gray, Beattie, and Collins. Grigg, Philadelphia

Hack M (1821) Harry Beaufoy; or, the pupil of nature. Harvey and Darton, London

Hack M (1834) Lectures at home. Darton and Harvey, London

Hack M (1839) Geological sketches, and glimpses of the ancient earth, 3rd edn. Harvey and Darton, London

Leach C (2006) Religion and rationality: Quaker women and science education 1790–1850. Hist Educ 35(1):69–90

Meyers G (1989) Science for women and children: the dialogue of popular science in the nineteenth century. In: Christie J, Shuttleworth S (eds) Nature transfigured, science and literature, 1700–1900. Manchester UP, Manchester, pp 171–200

Mingins R (2003) Light within or beacon without? An evaluation of the impact of evangelicalism on Quakers 1820–1840. Quaker Stud 8(1): 51–67. http://digitalcommons.georgefox.edu/quakerstudies/vol8/iss1/4/. Accessed 9 Mar 2017

O'Connor R (2013) The earth on show: fossils and the poetics of popular science, 1802–1856. Chicago UP, Chicago

Paley W (1821) Natural theology; or, evidences of the existence and attributes of the Deity, collected from the appearances of nature. Richardson and Co., London

The Annual Monitor for 1887 (1886) Obituary of the members of the Society of Friends in Great Britain and Ireland, for the year 1886. Samuel Harris and Co., Bishopsgate

The Athenaeum (1832) Geological sketches and glimpses of the ancient earth. No. 222:77

The Gentleman's Magazine (1832) A geological manual; principles of geology; sketches and glimpses of the ancient earth. 102:43–47

The Gentleman's Magazine (1853) The earth and its inhabitants by Margaret E. Darton. 39:70

Thomas AB (1912) The Beaconite controversy. Bull Friends' Hist Soc 4(2):70–81

Walker W (1843) Age—eight to eleven. The juvenile library; or descriptive catalogue of books. Walker, Manchester

Wood S (1829) Prayers for the use of Sunday schools with lists of books suitable to the children. Hunter, London

Chapter 5
Jane Kilby Welsh (1783–1853): Faith and Family

5.1 Biographical Overview

In most scholarship on women science writers, Jane Kilby Welsh (1783–1853) is relegated to a footnote or a brief mention, alongside the titles of one or two of her books, and perhaps a mention that she was once a student of Amos Eaton. Reconstructing the history of this lifetime native of Massachusetts and situating her writings within the larger context of both New England geology and theology as well as the familiar format has taken concerted effort and yielded limited, if not fascinating results, many of which are described here for the first time.

Jane Kilby Welsh was born in the Boston, Massachusetts area on January 25, 1783, to John Welsh Jr. (d. 1789) and Hannah Kilby (1763–1784). An only child, she was named after her maternal grandmother Jane Gridley Kilby (b. 1739) and nicknamed Jenny by her family. Either by general convention or typographical error, her last name is occasionally listed in both literary sources and official state records as Welch. She never married, and lived her entire life in Massachusetts[1].

Jane's childhood was marked by tragedy. Her mother died when she was a toddler, and her father passed away from a "lingering illness" (Welch 1902: 17) when she was only 6 years old. A Mason who served as Massachusetts Grand Recording Secretary, John Welsh was afforded burial with full Masonic honors. Jane's paternal grandfather, a shopkeeper, was executor of his son's estate, and it was determined that Jane would receive 15 £ out of the sale of her father's household property. When her grandfather died in 1812, he left her one-seventh of his estate, but there is no record of his playing any further role in her life (Welch 1902). Now orphaned, Jane's care fell to her eponymous maternal grandmother, who herself had suffered her share of loss. Jane Kilby's first husband, William Tyler Kilby (1736–1777) had

[1] Some of the vital facts in this section were taken from her gravestone, a picture of which can be found at https://www.findagrave.com/cgi-bin/fg.cgi?page=gr&GRid=57293003. Additional resources included digitized marriage, birth, and death records located through www.ancestry.com.

K. Larsen, *The Women Who Popularized Geology in the 19th Century*,
DOI 10.1007/978-3-319-64952-8_5

passed away when daughter Hannah was 14, and Hannah herself had passed on at the young age of 21. Jane Kilby remarried in either December 1788 or January 1789, to widower Elijah Hunt (1728–1802) of Northampton, and with the death of her father soon after Jane Kilby Welsh moved with her grandmother and step-grandfather to Northampton, where she was to remain for the rest of her life.

No information has been located concerning her early years in Northampton or her education, but she developed an interest in science that was fed by the public lectures of Amos Eaton. Although he became one of the most successful early nineteenth century popularizers and teachers of science in the Northeast U.S., science was not Eaton's original vocation. He gave up his law practice at the age of 40 and turned his mind to science. He first sought formal education in geology and mineralogy through listening to Benjamin Silliman's lectures at Yale (1815–16), and then became a proselytizer of science, traveling on foot across Connecticut, Massachusetts, and New York state through 1824 delivering popular lectures and lecture series as well as tutoring on geology and other branches of science (Merrill 1969).

During his travels Eaton carefully studied the local geological formations and published his observations and surveys. Eaton's public lectures were open to women, and it is known that Eaton lectured in Northampton and other local towns in 1817–18 (McAllister 1941). This is most likely where Welsh made Eaton's acquaintance. As noted in Sect. 8.1, Eaton played a major role in the education of three women who became pioneers in women's higher education, Emma Hart Willard and Almira Hart Lincoln Phelps of the Troy Female Seminary, and Mary Lyon, founder of the Mount Holyoke Seminary for Women. The last of these later continued her studies with Eaton by living with his family in Troy, New York after he took on the formal position that made him famous as a teacher of science at the Rensselaer School (Lutz 1964). This institute was established by New York landowner Stephen Van Rensselaer in order to train teachers. Eaton was responsible for the teaching of all the sciences, from botany and chemistry to geology, and stressed hands-on laboratory experiences for his students, a philosophy that resonates well with the familiar format (although Eaton did not himself utilize this style in his numerous textbooks and laboratory manuals).

For example, in his botany text *Botanical Exercises* (1820), he directs teachers to have students directly handle and observe plants at every class meeting. His pedagogical philosophy—which stressed field trips in order to gather samples and observe the natural world *in situ*, experimentation in the laboratory, and small seminars that emphasized student presentations rather than large passive lecture sections—was considered novel in its time, although today it is recognized as a standard practice in science teaching (Hilger 2008). Eaton's *Botanical Exercises* also included the following advice for teachers:

> I should not omit to mention a very meritorious little work, entitled a Botanical Catechism, published at Northampton, 1819. It contains 34 pages, written in the manner of question and answer, and embraces all the essential principles of practical Botany. It is the best book for very young students, particularly for ladies' schools, which has appeared. It was written by Miss Jane K. Welch [sic], of Northampton, Mass. (Eaton 1820: 4)

Jane Welsh had published the book, *A Botanical Catechism: containing introductory lessons for students in botany*, anonymously, the words *By a lady* appearing on the title page. While as Eaton noted, this work was slender in length, it was important and influential, Almira Lincoln Phelps later referring to it in her own *Familiar Lectures on Botany* (1829) as "the first attempt by an American lady to illustrate the science" of botany (Lincoln 1838: 234).

It is not known how Welsh's work originally came to the attention of Eaton, but in an August 7, 1819 letter Welsh asked Eaton for his opinion on whether she should attempt a work similar to her *Botanical Catechism* on the topics of Geology and Mineralogy. Eaton's reply no longer exists, but shortly after the publication of her two-volume work *Familiar Lessons in Mineralogy and Geology* Welsh wrote once again to Eaton, thanking her former tutor for his help and crediting him for awakening her initial interest in the science (McAllister 1941). The reason for the delay between Welsh's original idea for the book (1819) and its eventual publication (1832–3) is nowhere explained.

Jane died on September 12, 1853, the official cause of death listed as a fit, and was buried in Northampton. In her will (posted on July 31, 1849), Jane bequeaths $70 to

> Mr. Francis Alger of South Boston part of in trust however to be paid by him in sums of ten dollars each to Henry Codman Esq., President Edward Hitchcock, Mr. Thomas Nuttall, Mr. Josiah Holbrook and Professor Thomas Webster for the purchase of a book on something that will be acceptable to them as an acknowledgment though humble of his and their aid and encouragement in publication of my books on Mineralogy and Geology. (Massachusetts Probate Court 1971)

The reference is to several of the gentlemen of science who had written letters of support for these volumes, as will be explained in the following section.

5.2 Familiar Lessons in Mineralogy and Geology (1832–33)

Whatever else was occupying her time during the unexplained gap of over a decade between Welsh's letter to Eaton and the completion of her *Familiar Lessons in Mineralogy and Geology*, 1831–3 was a period of prolific publication, if not writing and editing. In addition to her geological works, she produced a story book *Simple Stories* (1832) and a second, more detailed work on botany, the anonymously published *The Pastime of Learning* (1831). While *A Botanical Catechism* was written in a question-answer format (as befit the work's title), *Pastime* was a full-blown conversation and more, in ways hearkening back to the more novelistic style of Maria Hack's Harry Beaufoy books.

Pastime's preface begins with the usual explanation to the audience of the benefits of learning science to their intellectual and spiritual growth. Welsh's first stated goal is "to impart, in a simple manner, and by familiar conversation, some preliminary knowledge upon Botany; a subject regarded by many as beyond the capacity of the youthful mind, or not sufficiently interesting – and unimportant as a branch of female education" (1831: 3). She argues against such claims, noting that in her experience children are often eager to engage in "a walk to collect specimens of this nature, with a view to the pleasure of investigating them" (Welsh 1831: 3). But there is more than scientific knowledge to be gained in the process, as

> everything which tends to fix the attention expands the mind; and that as the Physical powers are invigorated by a pleasant ramble, the Intellectual are strengthened, and the taste improved and elevated to a perception of the Beauties of Nature, which may ultimately lead to the most important of all pursuits – the Knowledge, Love and Reverence of their Divine Creator. (1831: 4)

The familial setting is well-established from the start; the work begins with Caroline noting to her mother, named Mrs. G, that she and her sisters had completed their regular "studies and recitations" and were waiting to go on a promised walk with their mother to begin learning botany (Welsh 1831:5). The female family members discuss the various flowers they see, and bring back some specimens to the house to examine more closely alongside their brother, Charles, during their evening family pastime (hence the title of the book). After this preliminary lesson, in which both parents take part, the author takes a step back, and in the next chapter gives an uncharacteristically lengthy and detailed back story of the family, including the individual ages, personalities, tastes, and limitations of the various children (who range in age from about 3 to 17). For example, 10-year-old Caroline

> was of a lively, buoyant character, with ardent feelings and affectionate disposition. Sincerity, and a total exemption from selfishness, and abhorrence of it and of anything like guile or deceit, were her strong characteristics. Of a quick capacity and lively imagination, she made rapid proficiency in her studies, when disposed to apply herself – which, we regret to say, was not always the case… [due to a] love of novelty… and from impatience, when she met with obstacles in her lessons difficult to overcome. (Welsh 1831: 45–6)

Great care is taken to establish the family as proper Christians; in fact, in a rather autobiographical twist, the eldest child, Anne, is actually Mr. G's niece, having been taken in after the deaths of both her parents. Mr. G's mother also resides with the extended family, who spend the winter in their Boston home and the summer in their home in the nearby countryside. It is also deemed important to explain that another child, Henry (who had been expected to be "rear[ed] for the ministry") had died before his sixth birthday of "that fatal malady, *water on the brain*" (Welsh 1831: 42; emphasis original). All these tragedies had seemingly strengthened the faith of this idealized family; indeed, the family custom for ending each day (at which the domestic servants were also present) was a Bible reading, "making brief and appropriate reflections which were suggested by the events of the past day, and closing with a concise and solemn prayer" (Welsh 1831: 29).

Both parents take an active role in educating the children, but as befits a work in the familiar format, the mother is clearly the main teacher (accompanying the children in their walks and directing them to observe nature as well as discuss it). An exception is the lesson on entomology, which is led by Mr. G While the description of domestic scenes and familiar relationships occupies a significant portion of the book, the conversations themselves often drift into a more lecture-type tone, with occasionally lengthy sections of scientific exposition punctuated by the children asking a brief question or making a brief explanation, apparently to break up the pages of text. It should be noted that Welsh cleverly self-references in the work. After a discussion of flowers, Mrs. G recommends to Caroline that she would "'find it of advantage to look over this preliminary treatise on the subject,' handing her a small book, called A Botanical Catechism, 'from which I formerly derived much information, and which has been highly recommended by professed botanists'" (Welsh 1831: 57–8).

While Welsh apparently felt comfortable enough with her background and status as an author in botany, the geological sciences were a different matter, perhaps explaining the lengthy time it took to ultimately produce these volumes. She explains in her preface to the first volume that the "liberal patronage awarded to a recent publication on Botany – not only by friends of the Author, but also by the Public – has induced her to prepare in a similar manner, and with the same motives, the following Exercises on Mineralogy and Geology," giving the unlikely impression that the lengthy first volume on Mineralogy was produced rather quickly after the publication of *Pastime* (Welsh 1832: 3). What is more likely is that this project had been in the works for some time (perhaps over a decade, if she had begun her work shortly after her exploratory letter to Eaton), and that she had held off finishing it and/or submitting it for publication until the success of *Pastime*. The work had also been subjected to peer review, and had been "sanctioned by the favorable opinion of gentlemen distinguished for their scientific and literary attainments, and extensively known, esteemed, and honored" (Welsh 1832: 3). These gentlemen she thanks by name, the same men that she later honored in her will, offering that she "is unable adequately to express her gratitude. By their examination of this production, their interest in its progress, and their wishes for its success; together with their recommendation of the best Authors upon the subjects, and their loan of several popular and valuable works, they have conferred favors upon the Author, which it is both a pleasure and a duty thus publicly to acknowledge" (Welsh 1832: 3–4).

She also stakes her claim to having the proper background to produce this work by listing the famous authors whose books she had consulted, the list reading like a Who's Who of nineteenth century science, including Europeans Robert Bakewell, William Conybeare, Georges Cuvier, William Buckland, Charles Lyell, and Carl Linnaeus, and Americans Edward Hitchcock, Benjamin Silliman, and Francis Alger. As in the case of *Pastime*, her stated goals are not only to increase the scientific knowledge of children and teachers, but "she also fosters the hope, that the youthful mind, while thus exploring the works of Nature, will be elevated with gratitude, adoration, and love, to the view of Nature's God" (Welsh 1832: 4).

Her publisher, Clapp and Hull, issued lengthy advertisements for the book that included quotations of support from some of these same scholars. For example, Edward Hitchcock noted "I am pleased with your work in general, and think that by interweaving domestic scenes with Natural History, you will attract more so its pages than by a naked exhibition of facts. I think it will do much to advance the cause of Natural History, of Virtue and Religion" (*The Boston Medical and Surgical Journal* 1832: 164). Harvard chemistry professor John W. Webster (1793–1850) offered "Many works, designed wholly for young ladies or beginners, are too learned. You have made Geology far less repulsive to young minds than any one who has preceded you. I beg you to be assured I feel a deep interest in your work. I wish you all the success that you can desire, and shall be happy to aid you if in my power" (*The Boston Medical and Surgical Journal* 1832). Francis Alger (1807–63), famed amateur mineralogist, proclaimed

> By blending, in the form of domestic scenes, lively moral and religious reflections, with naked facts and details in science, you have given a character to your work, which happily adapts it to the wants of *young students*; while its easy, familiar style, and conformity in arrangement with the latest and most approved systems, cannot fail to lay open to their minds, as well as to general readers, a competent knowledge of two of the most important branches of Natural History. (*The Boston Medical and Surgical Journal* 1832)

Less enthusiastic, yet still supportive, were the comments of Josiah Holbrook (1788–1854) of the New England lyceum movement, who offered "The plan, materials, and spirit of the work, I have no doubt, from the slight examination I have been able to give it, will render it an acceptable gift to the cause of Science and of Popular Education, and am therefore glad to learn that you have concluded to give it to the public" (*The Boston Medical and Surgical Journal* 1832).

Familiar Lessons is very similar to *Pastime* in both style and in using the same family, but interestingly it is set a few years later than the first work (although in reality they were only published a year apart). This is perhaps a plot device that Welsh thought necessary given the subject matter, with botany considered easier for younger children to grasp than the geological sciences. This aligns with Maria Hack's aging of Harry Beaufoy between her two novels (Sect. 4.2). Welsh explains to her reader that "The lapse of a few years had produced but little change in the pursuits and pleasures of the residents of Oak Grove [the White Cottage]. The characters of the younger members of Mr. G's family, however, as they were more fully developed, evinced progressive improvement, commensurate with their rapid advance to maturity" (1832: 9). She then catches her reader up on how the children have matured, for example noting that

> Charles, nearly at the age of seventeen, possessed the same noble disinterestedness for which he had formerly been distinguished; together with a degree of manly firmness and stability of conduct, which gave decision and dignity to his character, without rendering him assuming or overbearing.... Fondness for literature, and a thirst for knowledge, together with the *love of virtue* —which was early impressed upon his heart — were now the predominant traits in his character. (Welsh 1832: 9–10; emphasis original)

His sister Caroline "continued the same frank, cheerful, affectionate, and ingenuous girl as formerly; incapable of disguise herself, and with feelings repugnant to every species of dissembling in others" (Welsh 1832:10), while "Emma, on entering her thirteenth year, continued the gentle, sensitive being that she had always been," having a particular sympathy towards the plight of animals (Welsh 1832: 11–12). Julia, the youngest, was now 8 years old, old enough to have "been promised admittance to the library during the anticipated lessons, and a view of the minerals in her father's cabinet; with the injunction, however, not to misplace those she might be allowed to handle" (Welsh 1832:13). Anne, the eldest (Mr. G's orphaned niece), was now Mrs. Arlington, a doctor's wife with a daughter of her own, Emily. Dr. Arlington, Anne's husband, had a business in Philadelphia, and his wife, daughter, and mother spent some time with Mr. and Mrs. G each year.

Welsh takes particular care to establish Anne's good Christian character, and her devotion to her husband (including the required social duties of a doctor's wife) before explaining Anne's scientific credentials (the ranked order being vitally important), including the fact that she had gained much of this knowledge from her husband:

> Notwithstanding her many engrossing duties, with so much judgment was the distribution of Mrs. Arlington's time regulated, that she had it in her power to pursue her botanical researches much to the advantage of her husband, to whom her knowledge of this subject rendered her an important auxiliary in the *materia medica.* She had also obtained from him some knowledge of Mineralogy and Geology; upon which subjects he had acquired considerable knowledge, particularly during his absence in Europe. (Welsh 1832: 15)

Welsh also takes time to explain the geological prowess of the parents Mr. and Mrs. G, mirroring the attention she had paid in setting up her own expertise in the preface. Mr. G had originally been too busy to devote much time to geology, but more recently, "having more leisure, he had engaged with renewed pleasure in examining the recent publications upon the subject; and he had arranged his minerals in the *Geological Apartment* – the name by which a small room connected with the library was designated" (Welsh 1832: 16). Note that the G family has not merely a simple natural history cabinet, but an entire room dedicated to their collection. Mrs. G's geological knowledge is more problematic, perhaps in order to make the character properly deferential to her husband's presumed superior wisdom in the science. She recalls that before her marriage she had enjoyed attending (presumably public) lectures, "but the theories had since been so various, and so frequent were the changes in these branches of science within the last twelve years, that she was forced to acknowledge her ignorance, and must place herself among Mr. G's pupils" (Welsh 1832: 18). For his part, her husband graciously offers that she must have some up to date knowledge from her casual reading of current journals, or else she would not know that her knowledge was in need of updating; hence "he should look to her occasionally as an assistant" (Welsh 1832: 18).

One particular lesson led by Mrs. G involves a discussion on the health hazards of lead, for example when used in a pottery glaze. Her mother-in-law, Madam Arlington, muses that in this age of new discoveries a "poisonous principle is now found lurking, unseen, in some of the mineral substances which were formerly used with impunity, and, till apprised of their danger by some fatal result, regarded as harmless" (Welsh 1832: 273-4). Mrs. G uses this example as evidence that basic scientific knowledge needs to be imparted to members of "all classes of society," to which Anne adds that science needs to be "better adapted… to the more humble capacities of the weaker sex, as we females are designated; by whom they are so continually applied to use, for domestic purposes, in the daily occupations of life" (1832: 274). This illustrative conversation not only demonstrates the integration of everyday examples into the lessons, but issues of gender-related social roles and ethics. Another reference to the everyday experience of a Massachusetts family is a discussion of attending lectures at a local lyceum. One of Caroline's friends is a regular attendee, and Emma asks her father for permission for she and her sisters to do so when they return to the city at the end of the summer. Echoing Jane Marcet's impetus for writing her own *Conversations*, Mr. G explains that it was his intent that the family's lessons would adequately prepare all of them to attend the lectures, as "those who have some preliminary knowledge of the subjects upon which he lectures, will derive more advantages from this popular mode of instruction than the mere novice – and the pleasure they receive will also be enhanced by it" (Welsh 1832:132). The adults then take turns lauding Josiah Holbrook (who had given her work a cursory viewing before publication), invoking hyperbole in attributing to him the "universal establishment [of lyceums], by his unwearied exertions, in various parts of the union" (Welsh 1832: 133).

Commonplace examples from New England geology are often cited, including numerous examples of the work of Silliman and Hitchcock. When Caroline asks whether trap rocks are common in America, she is told of the range of features extending from New Haven, Connecticut through Massachusetts and Vermont, and that trap rocks also "according to Professor Hitchcock's observation, form the palisadoes on the Hudson, crossing the State of New Jersey to the Delaware…. Professor Silliman infers that their origin can be imputed to no agent but fire" (Welsh 1833: 196). Amos Eaton is also repeatedly referenced, and it is implied that Mr. G has personally spoken with the professor. Local minerals are also discussed, such as transparent quartz from Grafton, VT, Cyanite from Chesterfield, MA, Indianite from Bolton, MA, and Anthophyllite from Haddam, CT.

Hands-on experimentation with rock and mineral samples, as well as direct observations of geological structures, are central to the family's science lessons. The second volume, concentrating on geology rather than mineralogy, opens with a description of the family's habit of collecting minerals on their daily walks. In particular, young Julia eagerly seeks out attractive samples (with the aid of her brother and his rock hammer), collecting specimens of the "magnetic iron, tremolite, asbestus [sic], and idocrase, and a few of the other minerals which occur in the vicinity of Boston." She arranges these in her cabinet with Charles' aid (Welsh 1833: 10). Other than the rock hammer, technology is sparingly referenced within these volumes. Mr. G is initially reluctant to explain the use of a goniometer, thinking it too complicated for his audience, but relents, accompanied with a number of figures; however he does not actually let them try it out for themselves. Emma requests chemistry experiments to augment a particular lesson, but her father puts the responsibility for such demonstrations on his son Charles (to be conducted in the future and behind the scenes rather than included within the action in the book).

A discussion of calcareous deposits ends with the family taking a walk in order to directly observe some of these structures for themselves. As Welsh describes the experience,

> In thus taking a geological view of the hills and valleys, and the various strata that compose them, together with every prominent feature presented by nature in the country around, they became practically acquainted with the general appearance of most of those rocks in the mass, which constitute the primitive and transition formation. Even the little Julia could discriminate them. They returned invigorated by the exercise, and more enlightened upon the subject; together with a more indelible impression upon their memories of their former lessons, and with increasing gratitude for those high sources of happiness which are mercifully granted to us by Him who is the giver of 'every good and perfect gift.' (Welsh 1833: 165)

At the end of the second volume Mr. Arlington returns, and the family prepares to accompany the Arlington family back to Philadelphia, planning to visit a mine along the way. The geological structures they visit are summarized, as if to entice the reader to do the very same pilgrimage on their own:

> They observed at Weston, that sienite formed the uppermost rock; at Sudbury, granite; and at Marlborough, gneiss; - that at Shrewsbury and Worcester, gneiss and mica slate occur, and occasionally granite there.... They found the rocks at Northampton to be sienite and granite; and the Geology of every part of Massachusetts which they visited, to correspond with Professor Hitchcock's description, recently published,

likely a reference to his 1833 *Report on the Geology, Mineralogy, Botany, and Zoology of Massachusetts* (Welsh 1833: 388-9). The book concludes with the family visiting Mount Holyoke.

Direct observations and hands-on experiences of science are not the only common aspect of the familiar format that Welsh appears to stress; as is common in these works, religious references are interwoven throughout. Like Hack's works the chapters begin with quotations, although these have a definitively religious intent (but are not generally Biblical quotations). As in the case of other works described thus far, only representative examples of such references will be included. For example, after discussing the wonderfully fertile soil in the Connecticut River Valley Mr. G remarks "Alluvial depositions appear to be the agent employed by Nature—or with more propriety I may say, by Nature's God—to prepare the earth for the residence of the human race" (Welsh 1833: 290). Likewise Mrs. G proclaims that the study of minerals, in their diversity as well as common properties, could not

> fail to produce solemn and elevating reflections at the immensity of the productions of nature. 'That the Almighty Creator of *all* that constitutes this vast universe, both animate and inanimate,' she continued, 'should condescend to regard the concerns – relieve the wants – and promote the happiness of each individual, among this innumerable multitude; no less than if that one was the only being created, and the sole object of His providential care and love; cannot fail to impress the mind with wonder, reverence, and adoration.' (Welsh 1832:171–2)

Welsh utilizes a specific type of religious refrain throughout these volumes. At the end of each major section, Mr. G leads the family in a reflection upon the religious lessons of that set of topics. The volume on geology itself ends with such a three page religious reflection by various members of the family as to how their study has brought them to consider the "benevolent design and unwearied goodness" of God (Welsh 1833: 387). Mr. G personally notes,

> 'May the study we have pursued,' he added, 'apparently with pleasure to us all... be fraught with permanent benefit to the younger part of my audience, and be reverted to with pleasure by their seniors. Should it ultimately effect the most desirable of all objects — that of studying the Creator in His works and in His word —and conduce to the cherishing of a spirit of love and reverence for the divine perfections of the great Author of Nature, my highest aim will be accomplished.' (Welsh 1833: 387–8)

Finally, as is commonly included in such works, discussions centering on the relationship between geology and Genesis are included, including the Deluge and the six days of creation. However, unlike many similar works, Welsh includes a summary of

Penn's Mosaic geology (largely following Rennie's *Conversations on Geology*), brought up by Mrs. G, who, it will be remembered, was particularly knowledgeable about the changes in geological knowledge over the past decade, in the context of a discussion concerning the "strife between Neptune and Vulcan, as the opposing theories of Werner and Hutton are called" (Welsh 1833: 21). Mr. G willingly devotes three pages of discussion to this concept, although "every geologist regards his theory—as a respected friend of mine remarks, as 'among the wildest schemes that ever floated on the brain'.... it is presumed no geologist can be found in this country or in Europe who would advocate his theories" (Welsh 1833: 29). But whereas Maria Hack is careful to indicate direct quotations from other works and properly give credit in footnotes, Welsh follows the American convention (see Sect. 3.3) of not respecting copyright or the boundaries of intellectual property when it comes to foreign authors, liberally borrowing without attribution when it suits her purpose. Compare, for example, the following from Welsh's *Familiar Lessons in Mineralogy and Geology*:

> Mrs. G. remarked, she thought some of the many theories which had been proposed might be termed romantic. Even philosophers, when possessing a fertile imagination, have sometimes mistaken a fancy for a fact—and she had found it amusing to follow them. (Welsh 1833:17–8)

to the corresponding lines from Mrs. R in Rennie's *Conversations on Geology*:

> we must go through a great many facts and conjectures... beside some pretty romances, which are called Theories of the earth, and tell us how the world was made... philosophers, if they have much imagination, are apt to let it loose as well as other people, and in such cases are sometimes led to mistake a fancy for a fact. Geologists, in particular, have very frequently amused themselves in this way, and it is not a little amusing to follow them in their fancies and their waking dreams. (Rennie 1828: 5–6)

The derivative nature of Welsh's passage cannot be disputed.

But as Welsh has clearly demonstrated, she is capable of citing the arguments of others, when appealing to authority augments her argument. For example, a discussion of geologic strata (seen represented in geologic cross sections of Fig. 5.1) leads Mrs. G to bring up the fact that some interpret the word *day* in the Genesis to refer to "a period of several thousand years." Mr. G agrees, noting "Were this admitted, the difficulty of reconciling the creation with six days would be removed. Indeed, Mr. Bakewell suggests, that the six days in which the Creator called into existence different classes of beings, may imply six successive epochs of an indefinite duration" (Welsh 1833: 60). Mr. G continues further, positing a deeper connection between geology and theology: "Were any proofs wanting to authenticate the Mosaic records, Geology would supply them. Indeed, it has been remarked that the evidences of religion have advanced with the progress of true philosophy" (Welsh 1833:12). As in the case of Maria Hack, she distorts Cuvier's arguments (in this case his public lectures) while citing him by name (Sect. 4.2):

> Baron Cuvier remarks... that the ideas of Moses, respecting many of the highest points of Natural Philosophy, are proved by his books to have been perfect; and that, in a scientific view, his cosmogony may be regarded as remarkable—since the order of the different epochs of creation, represented by this sacred historian, precisely agrees with that which has been deduced from geological investigations. (Welsh 1833: 12–13)

Fig. 5.1 Geologic cross sections. Reproduced from Jane Kilby Welsh (1833) *Familiar Lessons in Mineralogy and Geology, v2*

As described in Chap. 1, while Cuvier did mention the Noachian Flood in his public lectures, (including the1805 lecture series at the Athenaeum in Paris and the Collége de France possibly referenced here), this was an expectation of geology lectures of his time. It should be said that Welsh was certainly not alone in her misinterpretation of Cuvier. Martin Rudwick (1997) affirms that Cuvier did point out a parallel between the order of fossils in the strata and the record in Genesis, but notes that this was not an original idea at the time. In addition, Rudwick (1997) recounts that reports published about these lectures suggest that many of the attendees also interpreted him as lending his support to a Biblical interpretation of the strata.

Another common misinterpretation of Cuvier's work is also offered, this time by the matronly Madam Arlington. After she voices the opinion that "the present appearances of the earth's surface… afforded evident indications of the deluge," Mr. G confirms that the "fact of a *universal* deluge, at no very remote period… is presented to the geologist in every point of view; and had we not been informed of the phenomenon by the sacred historian, we must have inferred that some great convulsion of this nature had agitated the crust of the earth" (Welsh 1833: 13; emphasis original). Here Welsh conflates several items. Firstly, Cuvier did not argue for a worldwide flood, only widespread events. Secondly, it is not clear here which of the supposed catastrophes Cuvier found reflected in the fossil record is to be accepted as the Noachian Flood. Such sweeping generalizations are difficult to disentangle, but nonetheless would have been accepted by readers with a predisposition to agree with such literalist sentiments.

The first volume of Welsh's work received a number of positive reviews. *The Boston Literary Magazine* (1833) wrote

> We have been much pleased with the perusal of this volume. It contains a scientific arrangement of the principles of these sciences, according to the latest and most approved systems, and is written in an engaging and spirited style. The interweaving of descriptions of lively domestic scenes and historical sketches, with the abstruse principles and forbidding *technicalities* of these studies, is such as to render the book much more attractive, particularly to the young, than would be a mere exhibition of naked facts. It contains sufficient information on the subjects of which it treats for the general reader, and needs only to be known in order to be extensively used. (343–4; emphasis original)

Likewise, *The Boston Medical and Surgical Journal* (1833: 196) offered of this "handsome duodecimo" that it

> contains lessons in Mineralogy and Conchology, written in a familiar style, suited for family use, and illustrated by about 80 engravings. The fair author of this manual has gathered information from sources the most authentic, and given it a form that cannot fail to be inviting to the mind of the young. Already well known, and having done a share of service to the rising generation by her illustrations of vegetable nature, she will fix still more firmly the foundation of her fame by these more profound investigations.

But praise was not universal for *Familiar Lessons in Mineralogy and Geology*, a fact that demonstrates not only the cloud under which authors in this genre were forced to write, but explains Welsh's lifelong loyalty to the gentlemen of science who had risked their reputations to provide words of support for her project.

5.3 Criticisms of the Familiar Format and Its Female Authors

In their nearly 20-page review of Amos Eaton's *Geological Textbook*: *prepared for popular lectures on North American geology* (1830) the anonymous critic or critics for *The North American Review* not only refer to Eaton's writings as "charlatanism in science" (1831: 483), and "one of the most extraordinary productions with which *pseudo-science* ever imposed upon the public" [emphasis original] (1831: 484), but also the source of "the evil consequences, that must result from the circulation of so much error amongst the ingenuous young men of the country" (1831: 490). Among their concrete charges are Amos' use of hyperbole and colloquial language and his propensity to not only create his own nomenclature, but his resulting "fickleness on account of the changes which appear in every successive book I publish; I confess this is the *ninth time* I have published a geological nomenclature, and that I made changes in each, of more or less importance" (*The North American Review* 1831: 486). To be fair, Eaton was a colorful and at times controversial figure in New England geology who did misinterpret some geological structures due to an inability to abandon the Wernerian doctrine (Spanagel 2014). But to his credit he did bring the geological features and rock types of New York and New England to the attention of European geologists, ultimately serving to advance the study of North American geology.

But does the vitriol of the critic help dissuade an impressionable public from purchasing a particular work, or merely satisfy the ego of the critic? In many ways the literary critics of journals of this time played the role of the Chef Gordon Ramsay of their day. Take, for instance, the following review of renowned Yale professor Benjamin Silliman's *Elements of Chemistry in the Order of the Lectures given in Yale College*:

> Our booksellers' shelves are already bending under their load of elementary works on chemistry, and we are very sorry to see that Mr. Silliman has added two ponderous octavos to the weight. We are sorry to see this; 1. On account of his own reputation. It will not be a gainer. 2. We are sorry to see this on account of the publisher. He will be a loser. 3. We are sorry to see this on account of the public. The public does not want it. (*The American Monthly Review* 1832b: 199)

In this case, the critic seems to find fault not with the author's scientific knowledge, but merely the format of presentation, suggesting that by the 1830s the market was already saturated with introductory level science books. By that decade the disdain of some critics clearly tipped against not only introductory texts in general, but the familiar format in particular. Recall that in their otherwise positive review of *Conversations for Children* by Jane Marcet *The London Quarterly Review* (1844: 12) had noted that it was written in "that garrulous form we so much condemn."

But such wholesale criticism of the format was not universal. On the contrary, in their review of Delvalle Lowry's *Conversations on Mineralogy*, *The Quarterly Journal* recommended that "every work, which, like Mrs. Marcet's, tends to accustom the young mind to a connected train of thinking, and by easy and familiar, yet not frivolous, means, brings it acquainted with the sciences, does almost incalculable good, and the sooner such works are put into the hands of young persons, after they have attained their ninth or tenth year, the better…." (1823: 155). It is undeniable that introductory works in general, particularly those written in the familiar format, were subject to increasing criticism as the century progressed, and there appear to have been several specific aspects of such writing that critics particularly found fault with. One was with the writers' credentials themselves. In her *Studies in Evolution and Biology* (1890) physician's wife Alice Bodington (1840–1897) addresses this prejudice head-on:

> A stigma is supposed to rest, for some mysterious reason, upon the person who ventures to write upon any branch of science without being an original discoverer. I am at a loss to imagine why it is considered almost wrong to write about physical science without having made original experiments. A historian is not required to have fought in the battles he describes, nor a geographer to have personally traversed the wilds of Africa. Why cannot a wide view be taken by some competent person of the results of the labours of hundreds of scientists, so that we may more clearly see what manner of fabric is being reared? (Bodington 1890: ix–x)

In response, an openly critical, anonymous (and presumed to be male) reviewer of her work in *The Selbourne Society's Magazine* countered

> But are the cases exactly parallel? Warriors and statesmen make history; scientific men are not expected to make the laws of nature, which they have but to record; and just as we expect of the historian an acquaintance with original documents, it seems not unreasonable to make a like demand of writers who undertake our instruction in this other department. The facts of nature are the original documents from which we have to draw. (1891: 229)

He cites as evidence of the harm that can come from a nonscientist writing about the natural world a rather flowery description of "the whole race of fungi, which instead of getting wholesome nourishment from the free air of heaven and the bosom of Mother Earth, are dependent upon organic material, animal and vegetable," stating that it "would hardly have been written by a practical naturalist" (*The Selbourne Society's Magazine* 1891: 229).

This leads to a second accusation that was (and often still is) lodged against popularizations in general, that the adaptation and simplification of science for a nontechnical audience leads to the unnecessary distortion *of* that science (Hilgartner 1990). For example, an anonymous reviewer of Delvalle Lowry's *Conversations on Mineralogy* notes that a "few puerilities now and then occur, which perhaps the plan of the work has almost unavoidably introduced, and may be some objection to the plan itself," because, as he argues, it implies that the reader is not intellectually sophisticated enough for "the comprehension of the more difficult parts of the subject" (*The Quarterly Journal* 1823:159). One of the specific examples cited by the reviewer is the simplified statement (meant to pique the reader's interest) that diamonds are essentially charcoal in a different state; in response, the reviewer recommends that Lowry revisit Marcet's *Conversations on Chemistry* in order to brush up on her chemistry.

In reviews of the familiar format the condemnation of the simplification of science can be particularly harsh. As surgeon J.S. Forsyth accuses in his work *The First Lines of Botany* (1827),

> The baby system of education, as old as Adam, called the 'Interrogative,' hitherto so slavishly enforced and patiently endured, as regards the higher order of the sciences, has nothing to recommend it, not even simplification of ideas; on the contrary, there is much to condemn in its indiscriminate application; and in most instances it is inferior, as a channel of communicating the elements of general knowledge, to the usual methods of plain, analytical, and rational display. To children of tender age, indeed, it may possess some advantages, when the subjects are suited to their years and capacity. In these cases it appears plausible enough to connect a string of monotonous laconisms, in the shape of continued question and answer; but where no distinctions are drawn between an infantile and a mature mind – the idea is as ridiculous as absurd. (1827: 15–16)

Similarly, in his review of Rosina M. Zornlin's *What is a Voltaic Battery*, Charles Morgan offers "To speak ill of a class of literature, the want of which had been so grievously felt, may seem disgracious and ungrateful; yet it may be doubted whether the scholar of other days did not get on better without nursery libraries, than the actual generation does now, amidst the overpowering multitude of ill-conceived and badly-executed works" (1842: 833). His main argument is that before works written for children and women were readily available,

> the student was forced upon manly reading; and as far as style, taste, and imagination are concerned, he had a better chance of being raised by such reading to the intellectual level of the educated, than by the shallow superficialities and good-child plausibilities which are now especially provided for his instruction…. If the didactic works addressed to the mature intellect conveyed but slowly their instruction to the minds unprepared for perusing them, that instruction was better calculated to develope [sic] the faculties, and form the understanding of youth, than the crude, ill-digested trash, administered at present under the notion of writing down to a limited intelligence. (Morgan 1842)

Having noted some of the general criticisms made against such works in general, it is appropriate to examine in detail a particularly scathing 20 page review of volume 1 of Jane Kilby Welsh's *Familiar Lessons in Mineralogy and Geology* that appeared in *The American Monthly Review* in December 1832.

The anonymous review begins by lamenting the fact that "There are no literary degrees for ladies" because

> It is a source of serious inconvenience to us. We therefore respectfully advise all female authors to affix numerals to their names, indicating their age. As things are, we have no means of deciding what may be the age or standing of a literary lady, who chooses to appear before us as an author or compiler. No combination of letters is appended to her name, which stand like the finger on a guide-board, to tell us by what road the traveler has come. We meet with females on the great highway, mingling in the scrub race of authors, and they expect of us the courtesy due their sex. We meet them so hooded and muffled, that we know not from their dress, whether they are misses in their teens, or maidens of fourscore. If, therefore, in our remarks on the work under review, we are wanting in that tenderness which we are disposed to show to the jejune effusions of yearlings, or in that severity which is due to those of riper age, we hope that we shall not be set down as ungallant gentlemen or dishonest critics. (*The American Monthly Review* 1832a: 482)

Having foisted the blame for the unpleasantness that is about to be perpetrated upon the author and her work unto societal conventions (and in a sense, for the author herself being unknown to them, ironically the standard practice in scientific peer review), he[2] then turns his attack to the well-known men of science who had endorsed her book:

> We know nothing of the lady before us.... She comes into our Star-chamber attended by a cloud of witnesses and defenders. We cannot however admit them in that character. We think that they are more in fault than the lady herself; at least they are accessories before the fact. They have urged forward this unfortunate woman to the perpetration of one of the greatest offences against the laws of the Commonwealth of Letters, that of putting forth a book on a subject about which we believe she is not only innocent, but utterly ignorant. Hear her defence. (1832a: 482)

With that final sarcastic comment, the attack begins in earnest, but not before impugning the reputation of some of these well-known scholars. The reviewer first exclaims with faux innocence that given "the mass of errors compiled by Jane Kilby Welsh... we cannot, without great violence to our better feelings, bring ourselves to the belief, that all of these gentlemen have 'examined this production'," but then adds "whilst, on the other hand, we find it equally easy to believe that *some* of them have," possibly referring to Holbrook (1832a: 483; emphasis added). The gentleman reviewer further proclaims that it is his reluctant duty to continue to review "this misshapen and monstrous production" although it is "too insignificant on its own account to require it" because "justice to the lady, to her friends, and above all to the public, demands of us such an examination as will sustain our accusation" (1832a: 483).

[2] The presumptive pronoun will be used, despite the possibly *Royal "we"* utilized in the review.

The reviewer then turns his weaponized pen to the literary style itself, the only part of the work that he considers the fault of the author herself. After rather sarcastically describing the characters and the detailed domestic scenes described in the first section of her work, he attacks the notion that Anne Arlington's botanical knowledge could possibly be of aid to her physician husband: "What complaint she is administered in, the author does not say. We have found her a dose. If the author had been a good '*herb-woman*,' we dare say she could have used Anne as a powerful sternutatory [sneeze inducer]. As she is, she is quite soporific" (1832a: 484; emphasis original). The conversation itself (what he terms the "repulsive face of dialogue" [*The American Monthly Review* 1832a: 484]) is described as

> one of the few portions which we can read, assured that we are not imbibing errors. Indeed, there is very little of womanly imagination in the conversation of the females who compose the audience, and such utter recklessness of all graceful and easy talking, that were it ever true that the study of the natural sciences can destroy these beauties in a female writer, we should hold up the author as a bright example of such lamentable effects. (1832a: 484–5)

He also condemns the heavy-handed preaching by Mr. G at the end of most chapters, advising the author "to lay to heart an old but quaint saying; 'Overdoing is undoing'" (1832a: 485).

The reviewer then deviates from his own original statement of only criticizing Welsh for her writing style, and attacks her rendering of the science proper, but couches it in terms of the chemical knowledge of her character, Mr. G, complaining "as he limits a knowledge of chemistry almost to its language, we are not at all surprised that he commits very serious mistakes, as often, almost, as he opens his mouth on chemical subjects" (1832a: 486). The first example is an overstatement of perhaps imprecise language (or awkward paraphrasing), in which Welsh describes an object's volume, here called the "bulk," reducing as the material is heated and some of "its particles evaporated" (1832a: 486). The reviewer also takes issue with Mr. G's somewhat nebulous use of terms such as the "heat of temperature" without definition (*The American Monthly Review* 1832a: 486). Another example is describing hydrogen as one of the "compound elements of water" (1832a: 487). Although it is clear from the context that she means that hydrogen is one of the constituent elements in the compound water, the reviewer is not satisfied. Another issue is the statement that "Water consists of oxygen and hydrogen, in the proportion of 88 ¼ of the former to 11 3/4 of the latter" (*The American Monthly Review* 1832a: 487). After a page and a half of over the top hand wringing in which alchemy, Greek mythology, and the Tower of Babel are invoked, it is revealed that the accepted ratios at that time were actually 88.9 and 11.1, an error certainly not deserving of such an extended tirade (1832a: 488). Likewise seven pages are devoted to criticizing the nomenclature and classification of acids utilized by Welsh.

Finally, several pages focus on supposed errors in the discussion of coal (especially in New England), in which Welsh had cited Thomas Nuttall by name. Noting that "No person stands higher in our affections, for his zeal for all departments of nature, for his sound and deep knowledge, and, not among the least traits, which we think conspicuous in his writings, for an intense love of truth, and for manly candor," the reviewer explains that Nuttall was contacted for a statement but he had "left the country," and an open invitation is offered to Nuttall for the purpose of defending himself (1832a: 500). Two non scientific errors are criticized, a comment about "Mr. Holbrook's apparatus, lately published" (the reviewer ridiculing the language by noting that you can't publish an apparatus), as well as the previously noted hyperbolic statement that Holbook deserves the honor of being called responsible for the "universal establishment" of lyceums, to which the reviewer notes that lyceums were operating in England and other parts of the US before Holbrook became involved (1832a: 501).

The reviewer then makes an aside comment about the section on conchology, but in the interest of the length of the review says nothing further on the topic, and instead begs Welsh to not publish the second volume of her work, before offering one parting piece of condescending advice: "Rely upon yourself. Trust not your friends. They have, we know not how it has happened, betrayed you. Trust not to the magic of a name.... Stand, therefore, gentle lady, in your second volume, on your own hard-earned reputation" (1832a: 502).

Welsh's second volume was published, the following year, but no review in *The American Monthly* has been located at this time, so it is unknown what sort of apoplectic state it might have generated in the reviewer. Perhaps the color plates might have offered them additional ammunition for critical analysis (Fig. 5.2). The close analysis of this review, as demeaning and unpleasant as it is, serves two useful purposes; specifically, it suggests why Welsh felt that she needed to acknowledge in her will the men of science who had supported her work, and more widely, it demonstrates the atmosphere in which these women published their works and the types of attacks to which they opened themselves when delivering these popularized works to the public eye. It may also explain why Jane Kilby Welsh never published another science book after completing this project. Her last work was a religious one, *A View of the Heathen World: And of Light Dispelling the Darkness. With questions, adapting it to Sabbath Schools and juvenile associations* (1834).

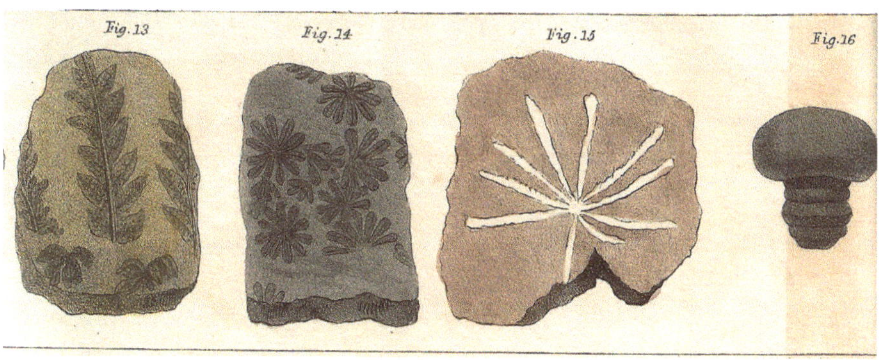

Fig. 5.2 Representative fossils. Reproduced from Jane Kilby Welsh (1833) *Familiar Lessons in Mineralogy and Geology, v2*

References

Bodington A (1890) Studies in evolution and biology. Stock, London
Eaton A (1820) Botanical exercises including directions, rules and descriptions, calculated to aid pupils in the analysis of plants. Websters and Skinners, Albany
Forsyth JS (1827) The first lines of botany. Bulcock, London
Hilgartner S (1990) The dominant view of popularization: conceptual problems, political uses. Soc Stud Sci 20:519–539
Hilger CM (2008) Almira Hart Lincoln Phelps. In: Patterson D (ed) Early American nature writers: a biographical encyclopedia. Greenwood, Westport, CT, pp 281–287
Lincoln AH (1838) Familiar lectures on botany, 7th edn. Huntington, New York
Lutz A (1964) Emma Willard: pioneer educator of American women. Beacon Press, Boston
Massachusetts Probate Court (1971) Probate records, 1660-1916. Probate Place, Hampshire, MA
McAllister EM (1941) Amos Eaton: scientist and educator 1776-1842. Pennsylvania UP, Philadelphia
Merrill GP (1969) The first one hundred years of American geology. Hafner, New York
Morgan C (1842) What is a voltaic battery? by Rosina M. Zornlin. The Athenaeum, no 778: 833–834
Rennie J (1828) Conversations on geology. Maunder, London
Spanagel DI (2014) DeWitt Clinton and Amos Eaton: geology and power in early New York. Johns Hopkins UP, Baltimore
The American Monthly Review (1832a) Familiar lessons in mineralogy and geology, designed for the use of young persons and lyceums, v1. 2(6):482–502
The American Monthly Review (1832b) Elements of chemistry in the order of the lectures given in Yale College. 1(3):199–209
The Boston Literary Magazine (1833) Familiar lessons in mineralogy and geology. 1:343–4
The Boston Medical and Surgical Journal (1832) Advertisements. New work on mineralogy and geology. 7(10):164
The London Quarterly Review (1844) Conversations on land and water by Jane Marcet. 147:12
The North American Journal (1831) Geological text book, by Amos Eaton, 1831. 32:471–490
The Quarterly Journal (1823) Conversations on mineralogy, with plates, engraved by Mr. and Miss Lowry, from original drawings. 14:154–160
The Selbourne Society's Magazine (1891) Short notices of books, nature notes. 2:229–232
Welch C, Welch A (1902) Welch Genealogy. Welch and Welch, np
Welsh JK (1831) The pastime of learning with sketches of rural scenes. Cottons and Barnard, Boston
Welsh JK (1832) Familiar lessons in mineralogy and geology, vol 1. Clapp and Hull, Boston
Welsh JK (1833) Familiar lessons in mineralogy and geology, vol 2. Clapp and Hull, Boston

Chapter 6
Delvalle Lowry (1800–59): Apprentice and Author

6.1 Biographical Overview

As is the case of many women writers of science, more is known about the more famous members of Delvalle Lowry's (1800–1859)[1] family than about the author herself. But of all the women described in this volume, Lowry perhaps most nearly lived in a gender-equitable environment in terms of science, as both her parents were well-respected in their field and she received an education in science that was in some ways superior to that afforded most men of her time. This chapter includes what is believed to be the most complete biography of Lowry attempted to date.[2]

Delvalle Eliza Rebekah Lowry was born on September 22, 1800 in Marybone, London. Her parents were famed engraver Wilson Lowry (1762–1824) and his second wife, mineralogist Rebekah Eliza Delvalle (1761–1848). Wilson Lowry was born in Cumberland to painter Strickland Lowry (1737–c.1780) and his wife Sarah Watson Lowry, and spent his childhood in both England and Ireland. Wilson inherited a love of the arts from his father, and at age 15 expressed an interest in becoming an engraver. He spent the next three years supporting himself by painting houses while he accumulated the necessary experience in both engraving and sketching. At age 18 he was hired by surgeon William Blizard and attended anatomical lectures. Although he was encouraged to complete medical training, he did not, but utilized what training he did complete when he subsequently specialized in engraving illustrations for scientific works (*The Annual Biography and Obituary* 1825). In addition

[1] Her year of death is often erroneously given as 1860.

[2] Thanks to the artistic talents of her family, three portraits of Delvalle are extant. The first is a pencil sketch by John Flaxman of Delvalle as a young teenager or preteen (http://stairgalleries.com/auction/06-25-11/catalogue/301-400/cat-331-340.htm). The National Portrait Gallery owns a pencil and watercolor portrait by her sister Matilda Heming dating to the 1830s (http://www.npg.org.uk/collections/search/person/mp06275/delvalle-elizabeth-rebecca-varley-nee-lowry), while her husband John Varley sketched her in pencil (http://www.artvalue.com/auctionresult--varley-john-i-1778-1842-united-portrait-study-of-delvalle-low-3359894.htm).

K. Larsen, *The Women Who Popularized Geology in the 19th Century*,
DOI 10.1007/978-3-319-64952-8_6

to engraving for Rees's *Cyclopaedia* for two decades, he maintained relationships with many men of science, who would have been known to members of his household, including Delvalle. He was especially knowledgeable in the geological sciences and was an original member of the Geological Society. At the time of his death it was said that "his scientifically arranged cabinet is surpassed by few private collections in London. Indeed, his opinion was constantly sought by professors of mineralogy…." (*The Annual Biography and Obituary* 1825: 101).

He also turned his scientific and technical interests to his craft, inventing a ruling machine around 1790 for the purpose of improving the accuracy of lines, and developed other instruments in 1801 and 1806 for making elliptical curves and perspective drawings. He was also "the first engraver who used diamond points for ruling, and he discovered the secret of biting in steel successfully" (O'Donoghue 1893b: 213). He was elected a fellow of the Royal Society in 1812. Lowry's first wife was a Miss Porter of Birmingham, the mother of his two eldest daughters, Anne and Matilda. Anne became the wife of Hugh Stuart Boyd (1781–1848), a Greek scholar who famously taught Elizabeth Barrett Browning near the end of his life, while he was blind (Hunt 1886). Matilda (1796–1855) became a successful artist in her own right, as a portrait painter. According to Ralph Peacock her work "is interesting in a rather more special way. It is distinctly rare to find the ordinary landscapist of her time working with an eye to truth rather than to the making of a so-called composition of the period, rare enough in fact to place her quite above the ordinary" (1905: 69). Matilda married Mr. T. Heming, a Magdalen College, Oxford, astronomer, who in 1814 had designed a pair of celestial hemispheres for demonstration and educational purposes that her father engraved (*The Literary Panorama* 1814).

Lowry wed for a second time on June 16, 1796, to Rebekah/Rebecca Eliza Delvalle (1761–1848). Rebekah came from a well-known family of Jewish merchants of Spanish descent; her parents were tobacco and snuff merchant Abraham Delvalle (1731–1785) and his wife Rebecca Henriques Sequeira (d. 1807), and her seven siblings included sister Abigail (1753–1801), who married stockbroker Abraham Ricardo, and gave birth to David Ricardo, a famous British political economist (Henderson 1997). Rebekah gave birth to two children, Delvalle and her younger brother Joseph Wilson Lowry (1803–1879). Joseph became an engraver, and worked in collaboration with his father and on his own works. He became interested in geology at a young age, and helped to construct a geological model for the Isle of Wight. He lent his engraving services to the Geological Survey of Great Britain and Ireland, and his plates of geological sections and fossils graced many of these documents (O'Donoghue 1893a). Joseph never married, and after his father's death in 1824 his mother came to live with him and became a tutor of languages. Rebekah Lowry was well-known in the scientific community, not only for her impressive mineral and fossil collection, but for her "considerable reputation as an instructress of mineralogy, and the elements of mathematics" (*The Annual Biography and Obituary* 1825: 107). She was widely sought after for private lessons, and Mary Somerville was perhaps her most famous student. John Jeremiah Bigsby, geologist and British Army doctor stationed in Canada between 1817–1825, was also one of her students (Wilson 2013). However, she was not only a teacher to the well-placed in society, and her lectures were open to all interested students.

Given her family background, it was inevitable that Delvalle would become proficient in geology, especially mineralogy, and would appreciate the need to learn from handling actual samples, utilizing hands-on techniques to study the properties of and identify various minerals. In her parents' home she also met some of the most famous geologists of her time, including Robert Bakewell, who had worked with Wilson Lowry to determine the hardnesses of a number of minerals (Bakewell 1819). Delvalle wrote three works of geology during her life, the first, *Conversations on Mineralogy*, was published in 1822 when she was merely 22 years old. It went through three British editions through 1837 and an American edition published only a few months after the initial British edition. The work featured plates engraved by Delvalle and her father from original sketches (Fig. 6.1). Her other works were published much later in life, after her marriage to artist John Varley (1778–1842), and were at least partially precipitated by her need to support herself due to her husband's infamous financial difficulties.

John Varley's biographer sums up the artist's life as follows: "It is given to but few men to fail so utterly in respect to everything relating to worldly affairs, and yet leave such a name behind both for goodness of heart and for the sterling results of genius, as did John Varley, the artist, and one of the founders of the Society of Painters in Water Colours" (Story 1894: 195). Despite an interest in and talent for drawing from an early age, Varley's father, Richard (d. 1791), "would not hear of his son's becoming an artist. Painting, said the elder Varley, was a poor trade, and none

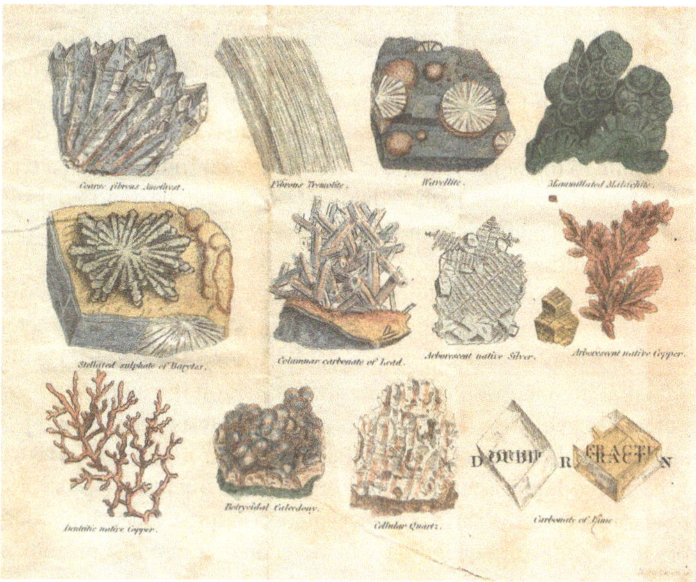

Fig. 6.1 Representative mineral specimens. Reproduced from Delvalle Lowry (1822) *Conversations on Mineralogy*, Amer. Ed

of his children should become artists" (Story 1894: 204). Instead, John was sent to a silversmith at age 13 to learn that trade. Upon his father's death less than a year later, the plans changed, and despite the fact that his mother, Richard Varley's second wife Hannah Fleetwood, was urged to find some other apprenticeship for him, after a failed stint with a law stationer John was allowed to become an artist (Story 1894). Three of his siblings followed suit: Cornelius Varley (1781–1873) became a watercolor painter and William Fleetwood Varley (1785–1856) was a landscape artist, as was their sister Elizabeth (1784–1864), who also married a painter, William Mulready. The three Mulready sons also became artists like their parents.

John established himself as a landscape painter in water colors; by age 19 he was exhibiting at the Royal Academy, and with his brother Cornelius helped to found the Water Colour Society. On a painting trip to Wales he had a run in with a bull, and during his later astrological studies Varley blamed the incident on his natal horoscope. Later in life he was attacked by a pack of hunting dogs and had his clothes ripped off (Story 1894). This interest in astrology came after his first marriage, at age 24, to Esther Gisbourne, whose brother John was in the same social circles as Shelley. Among his own social circles Varley became more famous as an astrologer than an artist, although there is no evidence he took money for the horoscopes he drew for his friends and acquaintances (Story 1894). Among his literary friends with whom he discussed astrology were Sir Richard Burton, Edward Bulwer-Lytton, and William Blake. He was also a friend of Wilson Lowry and lived on the same street, in a house/gallery in which he gave art lessons and occasionally also had students living with him.

It might have been better for Varley if he had become a commercial astrologer, as he had financial problems for most of his adult life (something he had apparently not seen presaged in his own natal chart). Esther bore him eight children—Albert Fleetwood, Henry, Frank (who died as a child), Charles Smith, Haydon, Emma, Susan, and Esther—and supporting this large family through his painting was a challenge (Story 1894). Even more so was an often demonstrated lack of sophistication when it came to financial matters, and he went bankrupt more than once, spending time in debtor's prison. Esther died in 1824, and he married Delvalle, the daughter of his friend and neighbor Wilson Lowry, the next year. The marriage did not produce any children.

By all accounts the two could not have made a more strikingly dissimilar pair. Twenty-two years his wife's senior, Varley was an "eccentric character" whose power of conversation was strikingly persuasive (Story 1894: 269). He also carried cure-all pills that he openly shared with others. She, on the other hand, was gentle, "affectionate and devoted to the last degree, and so far as her husband put it in her power to do so, made his later years happy and comfortable. All who knew Mrs. Varley spoke in the highest terms of her many good qualities" (Story 1894: 277). They were also opposites in terms of physical appearance. According to his biographer, Varley had "grand dimensions… which, when in a stooping posture, suggested to beholders the rear-view of an elephant" and was "like a farmer in appearance; generally somewhat untidy in his dress, wearing at all times and in all seasons a tail-coat with great salt-box pockets, a not uncommon habit in those days" (Story

1894: 275). In contrast, Delvalle was "as thin and lath-like as he was broad and stout, and many were the jokes passed at their expense" (Story 1894: 277). It was also said that Varley and his students would "toss Mrs. Varley from one to the other across the table" for amusement (Monkhouse 1899: 151).

In his later years, Varley invested a tremendous amount of money (both his own and some that he had borrowed) trying to perfect an 8-wheeled wagon, which led him into debt. He spent the last six weeks of his life sick and in hiding from his lenders in the house of his dealer, William Vokins, before dying of a kidney infection (Story 1894). Upon Varley's death, Delvalle received a pension from The Royal Academy's charitable funds (Hamilton 2015) and, according to Varley's biographer, the architect and engraver Thomas Talbot Bury (1809–1877) wished to marry Delvalle, but presumably she turned him down, as there is no record of a marriage (Story 1894).

Delvalle's later writings, *The Engineer's Manual of Mineralogy and Geology* (1846) and *Rudimentary Mineralogy* (1848) were written under her married name, and served a very different audience than her earlier *Conversations on Mineralogy*. *The Engineer's Manual* was solidly packed with technical information, and meant for the practitioner (presumably male). For example, the preface notes

> For want of this knowledge many disappointments have been incurred, and much money has been wasted by enterprising individuals, who have sunk mines to find metals and coals, which did not exist within a hundred miles of the locality so explored.... The object of this present work is to enable engineers and others, whose occupation gives them frequent opportunities of examining rocks and strata, to ... assist them, by a description of the external characters of minerals, to recognize them when they occur, and also to furnish them with some data on which to found their expectations of meeting with such or such minerals, according to the nature of the strata through which it may be their business to cut or excavate. (Varley 1846: 2–4)

Rudimentary Mineralogy, for the use of beginners; or guide to those who are in search of the valuable metals essential in practical science was also intended for a more technical (male) audience and was included as part of Weale's Rudimentary Series. John Weale was known as a publisher of inexpensive educational works and "did much for technical education in England" through his Rudimentary Series and Educational Series of books (Carlyle 1899: 89). *Rudimentary Mineralogy* went through at least four editions (1848, 1852, 1856, and 1859), the last two of which titled *Rudimentary Treatise on Mineralogy* and included "A Treatise on Rocks or Mineral Aggregates" penned by American geologist James Dana. Delvalle Lowry Varley passed away on December 23, 1859 in Pancras, London, at the age of 59 (*The Spectator* 1859: 20).

6.2 Conversations on Mineralogy (1822)

In many respects, Delvalle Lowry's *Conversations on Mineralogy* follows the traditional pattern established in Sect. 2.2, including a reference to Marcet's *Conversations on Chemistry* in her preface. But in several aspects, her work is singular. The first is the fact that she was only 22 years old when the work was published. Second, understanding the controversy concerning the myriad classification schemes present in mineralogy and feeling unsatisfied with the general scientific landscape, she somewhat boldly offers her own system in an appendix, but admits that "it is, perhaps, as defective as those which I have examined, and rejected because I was dissatisfied with them; for I am far from presuming that I can effect what has hitherto baffled the attempts of the most able Mineralogists and Chemists" (Lowry 1822: v). Third, unlike all of the other texts examined thus far, the work is uncharacteristically lacking in religious references. This may be due, in part, to her upbringing as the daughter of a Jewish mother who left her faith in order to marry a non-Jew. Her father's geological and religious views may have also played some role, although Wilson Lowry was neither an atheist nor a Biblical literalist. His obituary notes that he was

> neither a complete Wernerian, nor a complete Huttonian. He believed that both fire and water had contributed to the formation of the earth's surface; but the more he read and reflected on the subject, the more he seemed inclined to ascribe the larger share in that operation to fire. … that the formation of the whole [strata], from the first granite rocks down to the marl and gravel beds, could not have taken up less than a million of years; and that none of the various strata could have been deposited by the Mosaic deluge; at least none but the most alluvial soils. (*The Annual Biography* 1825:103)

In terms of aligning geology and religion,

> Mr. Lowry used to say that it was not fair or reasonable to expect that the Bible should contain an accurate philosophical system; that infidels ought not to attack the Scripture, because it did not contain such a system; that every one knew that the language of Scripture was highly figurative, and therefore that it was by no means difficult to believe with Bishop Horsley and other eminent men, that the six days of Moses were periods of unlimited duration. (*The Annual Biography* 1825: 103)

Such views were considered decidedly mainstream in his time, so it would not have been unreasonable for Delvalle to follow a similar theological interpretation of the natural world. Given that the main topic of her book was mineralogy and not geology, it was relatively easy for her to avoid the entire debate over stratigraphy and the Noachian flood, but it does make her work stand out in a very noticeable way. For comparison, Jane Kilby Welsh's volume on mineralogy (Sect. 5.2) was replete with references to various mineral deposits as evidence of God's benevolence and wisdom. A definitive explanation for the surprisingly secular nature of Delvalle's work has yet to be determined.

Delvalle's usage of fictional conversation is also unique, and best described as utilitarian. As she explains in her preface, "The method of communicating knowledge through the medium of conversations appeared particularly well calculated for such works …. it is hoped that the present undertaking, on a somewhat similar plan, may not be found unworthy of the attention of the public" (Lowry 1822: iii–iv). She is honest in her description of what she hopes to accomplish, which is "to give my readers clear ideas of the principles of Mineralogy and Crystallography" (Lowry 1822: iv). Again, it is noteworthy that this is her single stated purpose for this book, in contrast with similar works written by the women writers under consideration that include overtly religious goals in their introductions or prefaces.

Adaptations made for the benefit of her audience are clearly spelled out in the introduction, as well as what she considers to be the limitations of her volume. For example, she explains that she had decided to leave out the then recently offered Mohs system of Crystallography

> for two reasons; firstly, because it is merely a system of description, and does not in any degree explain the structure or formation of Crystals; secondly, because it is so abstruse, that, without entering into it minutely, I fear I could not make it intelligible to my Readers; and even were I able to do so, I think they would not find a knowledge of it so practically useful, as that of Haüy's theory

which she includes (Lowry 1822: iv–v). She also states that she has kept the various mineral synonyms and technical terms to a minimum (defining them when necessary), and instead includes a catalogue of mineral names (along with their Greek, Latin, and German etymology) as an appendix, something that she claims has not yet "been attempted in our language (I understand that a very complete work of this kind has been published *in German* by Professor Mohs, but I am not acquainted with it)" (Lowry 1822: vi; emphasis original). Finally, she also understands that her descriptions of minerals can hardly be original, but instead focuses on providing useful information on the more important (abundant and economically useful) minerals, and has "seldom neglected to mention the purposes to which they are applied in the different arts and manufactures, or any other interesting circumstances connected with them" (Lowry 1822: v). This focus on practicality and providing an exposure to the minerals most likely to be included in both natural history cabinets and relevant to everyday life are running themes in Lowry's work.

In her acknowledgments we see the influence of her parents through whom it is presumed that she was able to make the acquaintance of these gentlemen. Among these are mineralogist and mineral dealer John Henry Heuland (1778–1856) for help with German and "access to his splendid collection" (Lowry 1822: vi); William Phillips (1775–1828), author of *An Elementary Introduction to the Knowledge of Mineralogy* (1816) and founding member of the Geological Society of London "for his great liberality in allowing me to make use of his large collection of original drawings and measurements of crystals" (Lowry 1822: vi); mineralogist John Mawe (1764–1829) "who has frequently gratified me with the use of rare specimens and curious crystals" (Lowry 1822: vi); and Charles Dietrich Eberhard König (1774–1851) the keeper of the geology and mineralogy collections at the British Museum. Note that unlike the case of Jane Kilby Welsh, none of these gentlemen are thanked for help with reviewing the manuscript itself, and there is no evidence that they provided letters of support or introduction to her publisher. It is assumed that it was instead the reputation of her parents that was sufficient to convince the well-known London publishers Longmans, Hurst, Rees, Orme, and Brown to bring her book to the public.

As in all of the works in the familiar format, an emphasis is made on hands-on learning and direct observation of specimens. In the case of minerals, this is especially important. Fortunately, not only were natural history cabinets with mineral samples relatively common, but there existed a number of dealers from whom excellent samples could be easily obtained, as well as wooden models of crystals,

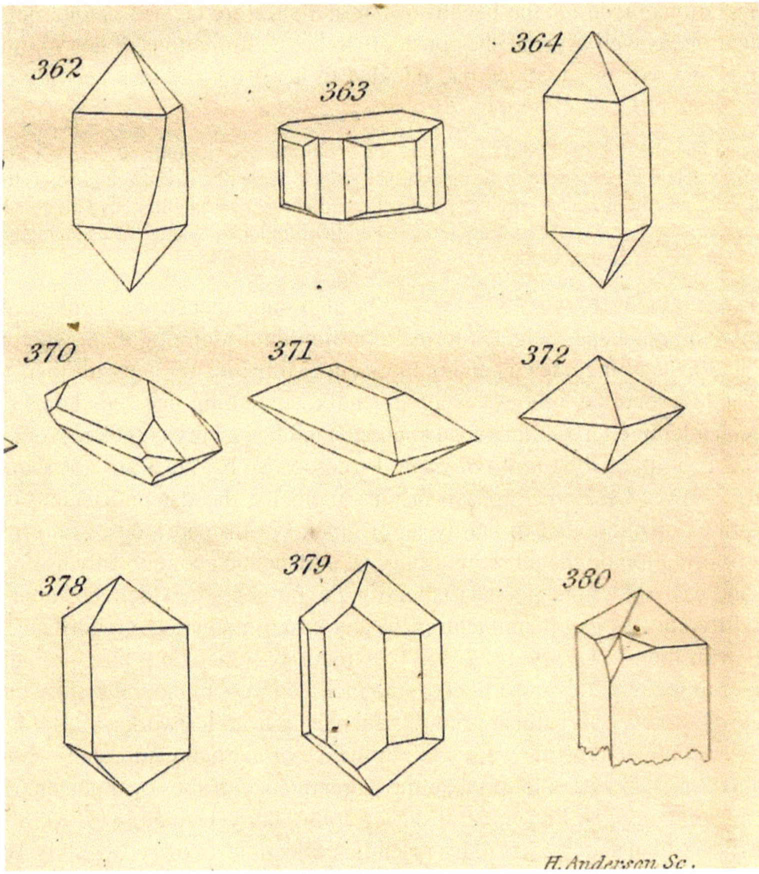

Fig. 6.2 Examples of crystal forms. Reproduced from Delvalle Lowry (1822) *Conversations on Mineralogy,* Amer. Ed

used to study their geometrical forms (Fig. 6.2). Lowry suggests several of these in her preface, including John Mawe and John Heuland. She is also careful to point out that her volume is only meant to "*prepare* the young Mineralogist for the study of more learned treatises" and warns that "very little knowledge of Mineralogy can be obtained from books, without an acquaintance with Minerals themselves" (Lowry 1822: vii; emphasis original). Lowry therefore recommends that her reader seek personal instruction, for example

> the private instruction given by Mrs. Lowry, (at her house in Titchfield Street,) who has for several years devoted a great part of her time to the completion of a very extensive and valuable collection of Minerals. This circumstance, together with her knowledge of other branches of science, renders her lessons at once interesting and advantageous to her pupils; and more particularly so to ladies, who, since their attention has been invited to scientific subjects, by the establishment of public institutions equally fitted for the instruction of both sexes, have cultivated Mineralogy with great ardour. (Lowry 1822: vii)

Note that not only is Lowry giving due attention to her mother's achievements as a teacher, but reminds her reader that public lectures are open to women as well as men.

The work begins with a domestic scene in which sisters Frances and Mary arrive an hour late for their lesson with their tutor Mrs. L (an obvious nod to Rebekah Lowry). Their excuse is that they had been enjoying the British Museum, with Mary noting that she had been particularly interested in the minerals, marvelling that these beautiful objects could be "natural productions" (Lowry 1822:13). Mrs. L asks them to recall some examples they remember seeing, and Mary mentions malachite. Mrs. L astounds Frances by explaining that malachite is mostly made of copper and that diamonds are "nothing more than charcoal" (Lowry 1822:14). Note the oversimplification that was referenced in Sect. 5.3. However scientifically sloppy this explanation might be, it is all that is needed to pique the girls' interest, and Mrs. L draws upon a familiar domestic metaphor in her explanation:

> Bread is made of flour, yest [sic], salt and water: yet we are not able to convert a loaf into these ingredients. So, though chemical experiments have proved that diamonds and charcoal are essentially composed of a substance called carbon, chemistry has not discovered any operation by which charcoal may be rendered pure and crystalline. (Lowry 1822: 14)

The discussion on diamonds leads them to discuss crystals (as seen in Mrs. L's ring) and Frances asks the question that sets the rest of the text into motion: "Do all minerals crystallize?" Mrs. L replies "No: I could tell you the names of a great many minerals which do not; but as you do not understand mineralogy, it would be useless to you to know them. However, if you wish to learn it, I shall be very happy to give you some instruction" (Lowry 1822:15).

As Lowry sets up the reader to expect in her preface, numerous references to everyday materials occur, again, something customary in this genre. Among these are slates used to draw on, plaster of Paris, and the purple tint in some windows caused by manganese. Mrs. L recounts common superstitions surrounding minerals as well, including the use of cat's eye "by the East Indians, who consider it as an amulet against bad eyes" and the belief that amethyst protects against intoxication (Lowry 1822: 85). The girls also engage in experiments with Mrs. L, including the common example of throwing a small sample of potassium into a basin of water and observing the resulting flash of light. They are instructed in how to identify minerals by their physical properties as well as the use of technology, such as the hydrostatic balance, electrometer, blowpipe, and goniometer. The girls not only learn how to use a common and reflective goniometer but are allowed to measure the angles of a crystal (unlike in Jane Kilby Welsh's *Familiar Lessons in Mineralogy and Geology* [Sect. 5.2]).

The ultimate goal is for the reader to be able to identify samples that they have not seen before by applying what they have learned, a process that Lowry has her characters clearly model. In Conversation XIV Mrs. L hands the girls a sample containing various minerals and asks them to identify what they observe. Frances easily sees that "Here are quartz and mica again; and I think these cubes are fluor [sic]: and here is pyrites, but I do not know which kind." Mrs. L asks Mary to make the identification, which she correctly does: "I see no foliated fracture, and where there is some indication of a regular form, the angles are certainly not those of a cube; it must be copper pyrites" (Lowry 1822: 258).

The book ends with the girls inquiring about learning geology. Mrs. L explains that geology cannot be learned via a cabinet, but "you must not only observe the different rocks and strata in various countries, but must also carefully compare those of the same kind occurring in different situations, or you will not be able to derive much advantage from books" (Lowry 1822: 300). Frances inquires if she should keep a journal of all the minerals and rocks she sees on a trip to Cornwall, but her tutor explains that she is likely to see complex mixtures that are beyond her cursory understanding, and instead if she is interested she should read William Phillips and William Conybeare's well-known *Outlines of the Geology of England and Wales* (1822), "an excellent and well arranged collection of geological facts; and Curvier's [sic] 'Introduction to the Theory of the Earth,' though it may be styled an elementary one, is truly scientific" (Lowry 1822: 301). Again, it is interesting that she does not take advantage of the mention of Cuvier's work to discuss religious connotations of geology, something that commonly occurs in other works of this type (e.g. the books by Welsh, Hack, and others). She ends the book by offering to share her "extensive collection of rock-specimens" with the girls at their return, if they are so interested, and tutor them in the subject (Lowry 1822: 301). Perhaps a sequel was initially considered, or at least the door was left open for a companion volume, although apparently none was ever written.

Conversations on Mineralogy enjoyed some success, and was considered sufficiently desirable to warrant an expanded and revised second edition (printed in two volumes in 1826), with additional sections on nitrates and chrome, among other materials, as well as a third edition in 1837. However, the claim in her father's obituary that it was "esteemed among the best works of its kind" was probably hyperbole (*The Atheneum* 1824: 192). A similar overstatement appeared in a review by *The Monthly Review*, which decided to "unhesitatingly characterize this Work as one of the most desirable Text Books that have issued from the British Press," a statement that was quickly picked up by her publisher for use in ads (*The Literary Gazette* 1823: 208).

An enthusiastic appraisal also appears in a letter of recommendation from John Redman Coxe, MD, that was included in the American edition of the book (also published in 1822) which reads in part:

> I can truly say, that, in my opinion, few works are better calculated to excite in young persons, a love for a science, which, although comparatively low in this country; has already established its vast importance to America, in a national point of view; and which is daily augmenting the number of its admirers. The conciseness of the work in question, and the plan adopted by its ingenious author, admirably fits it for the instruction of beginners; - nor is it wanting in interest to the more advanced lover of the science. The numerous plates from the graver of the celebrated Mr. Lowry, give it additional importance. (Lowry 1822: ii)

It should be noted, however, that the title page of the work itself clearly gives both Wilson Lowry and his daughter equal credit for the engravings (which were done from original sketches).

Judit Brody (1985), one of the few modern scholars to mention this work, criticizes Lowry for failing to write an engaging conversation that is worthy of the familiar format style of Jane Marcet, a criticism that is not without merit. Perhaps due to her rather atypically egalitarian scientific household, Lowry eschews all of the feminine subtleties of the familiar format. Very little time is spent setting up the personalities of the characters or the setting of the story, and one of the few hints of personality is Frances' love for crystals. The conversation style is short and succinct, with very few long expositional passages. The result is a stilted dialogue that is even less realistic than most of the genre, as evidenced in the following lines ending a discussion of mica:

> *Frances*. And what is the specific gravity of these substances.
> *Mrs. L.* That of mica varies from 2.6 to 2.79—the specific gravity of lepidolite is between 2.8 and 2.9.
> *Mary.* I suppose there are no more species in this family.
> *Mrs. L.* No; now we come to the slates. (Lowry 1822: 136)

On the other hand, an anonymous reviewer of Lowry's time called her style "easy, clear, and unaffected" (*The Quarterly Journal* 1823: 159) and *The Monthly Review* (qtd. in *The Literary Gazette* 1823: 208) termed her work "happily conceived, and it is executed with ability and taste," lauding the fact that the "author has studiously avoided all unnecessary parade of technical Diction."

However, the former of these reviewers was not entirely satisfied with the work, summarizing it as follows:

> On the whole, the *Conversations on Mineralogy* have strong claims to our praise; they contain a great deal of valuable information, delivered in a very pleasing language. The work is often enlivened by descriptions of the uses to which many of the minerals are applied, both in the arts and the common purposes of domestic economy, and cannot fail to stimulate the young mineralogist, who studies it with the attention it deserves, to pursue science with ardour and success. (1823: 160)

Such a description suggests that Lowry's work is a fairly representative (and successful) example of its genre. But as previously noted (Sect. 5.3), this same reviewer took Lowry to task for her oversimplification of the relationship between diamond and charcoal. The reviewer also notes errors of the scientific and philological variety. Lowry was apparently unfamiliar with the fact that wodanium, a supposedly new element first described in 1818 by Wilhelm Lampadius (1772–1842), had been shown to be a spurious result by Friedrich Stromeyer two years later. However, to a non chemist like Lowry the issue could still be seen as open in 1823 (the time that the review was written), as Lampadius was still referencing the would-be element in his scientific papers (Fontani et al. 2014). This reviewer appears to have a special dislike for wodanium, as he brings it up again in another context:

> The third conversation treats of specific gravity, of the hydrostatic balance, of the metals that are, and one that is not (*wodanium*) of their comparative utility, and of the external and physical characters of minerals. This is a pleasing and instructive conversation, from which the young reader cannot fail to derive a great deal of valuable information. We wish wodanium had been omitted. (*The Quarterly Journal* 1823: 156–7; emphasis original)

The reviewer also complains that there are a number of etymological errors in terms of mineral names, both in the appendix and the body of the text, which were noted "solely with the hope of seeing them corrected in a future edition" as they "are trifling mistakes, and easily corrected" (*The Quarterly Journal* 1823: 160).

As previously noted, there are clear differences between this early work, intended for younger beginners, and her later two books, written for a more serious and mature (although also introductory) audience. A clear example in the difference in style can be found by comparing the equivalent sections. For example, in *Conversations on Mineralogy* the girls and Mrs. L discuss the curious optical properties of Iolite:

> *Mrs L.*… Here is another substance that has also a very curious effect on light.
>
> *Frances:* Is it not blue tourmaline?
>
> *Mrs. L.* No; it is a much less abundant mineral, called iolite, or dichroite. In general it appears of a muddy violet, or blue indigo; but, if you look through it in the direction of the axis of the crystals, the colour is very different.
>
> *Mary:* That is very surprising.
>
> *Frances:* In this direction it is pale yellowish brown.
>
> *Mrs. L.* I have some small crystals imbedded in a white mineral, apparently albite, from Massachusetts—some of them are light blue, and nearly transparent…. (Lowry 1822: 111–2)

Note the connection to a specific place as well as the identification of another mineral that might be confused with it. The same discussion in her *Rudimentary Treatise on Mineralogy* is much more terse and technical in its language:

> 12. Iolite (Dichroïte; Indicolite; Cordierite; Peolom; Steinheilite).
>
> The property of dichroïsm was first observed in this mineral; and as it was then thought to be peculiar to it, it obtained the name dichroïte. It is rarely crystallized in six and twelve sided prisms; generally disseminated in small masses of irregular form, which are of a deep violet-blue when viewed parallel to the axis of crystallization; but in a contrary direction they appear yellowish or smoky grey. (Varley 1859: 49)

Hence it is clear that Delvalle Lowry, like Jane Marcet, was able to adjust her diction and effectively write works to serve different audiences depending on their needs.

6.3 Informal Apprenticeships and the Important Influence of Mentors

As described in Sect. 1.4, cultural norms severely curtailed the ability of nineteenth century women to gain field experience in geology. It was only considered practical and proper for them to walk either on their family lands or in their neighborhoods, or to accompany a male relative further afield (and only then in a location where the terrain would not be too strenuous for the presumed delicate female constitution). Studies such as mineralogy and astronomy afforded more freedom, not only physically but practically, because both could be done in the home. This was certainly true in the case of the Lowry women, especially Rebekah, who famously and very publically taught out of her home. The success of women in astronomy during this time period is relevant to the current discussion, as it demonstrates the importance of family in providing both training for women in science as well as acceptable roles as assistants.[3]

As Londa Schiebinger (1989) uncovered, in the seventeenth and eighteenth centuries a significant number of women worked in family observatories in Germany. For example, Johannes Hevelius, a wealthy Danzig engraver and amateur astronomer, built a private observatory across the roofs of three adjoining houses, including his own, for the purpose of observing stars with the unaided eye. Following the deaths of several assistants, his beautiful (and much younger) second wife, Elisabeth Korpmann (Koopman) (1646–1697?), became his assistant. After ten years of fruitful collaboration, their observatory (as well as much of the city) was destroyed in a fire in 1679. Hevelius is commonly said to have died of a broken heart eight years later, but Elisabeth oversaw the eventual publication of their work. During his lifetime, Hevelius immortalized his wife/assistant in a plate in his *Machinae Celestae*, where they are both shown observing with a brass sextant. This artistic depiction of a woman as an active astronomical observer clearly demonstrates the important role Elisabeth Hevelius played in their observatory.

An interesting but early parallel to Delvalle Lowry was Maria Clara Eimmart (1676–1707), daughter of George Christophe Eimmart, a Nuremberg engraver, painter, and amateur astronomer. Maria's illustrations (done at the telescope) graced her father's *Micrographia Stellarum Phases Lunae Ultra 300*. She married Johann Heinrich Müller, her father's assistant engraver, who shared the family's love of astronomy. Müller became a professor at Altorf, and was "aided by his talented wife" in his depictions of astronomical phenomena (Davis 1898: 137).

But not all familial apprenticeships ended happily. Maria Margaretha Winkelmann Kirch (1670–1720), the daughter of a minister of the Gospel in Panitzch (near Leipzig), learned both observational and computational astronomy from Christoph Arnold, a well-known amateur. Just as Delvalle had met her much older husband through the man to whom she was, in a sense, apprenticed (her father), Arnold introduced Maria to Gottfried Kirch, Germany's leading astronomer, a man some thirty years her senior. She became Kirch's second wife in 1692. It was only due to this relationship that Maria was able to continue with both her studies and astronomical work in their home-based observatory. Her work as her husband's assistant included calculations for calendars, observations of aurora, and the discovery of a comet in 1702 (Smith 1931). Ironically, it was the astronomical success of this husband and wife team that was to ultimately lead to Maria's professional frustration.

When Kirch became the official astronomer of the Berlin Observatory his observations moved from the attic of the family home to a public institution, which signaled the predictable decline of Winkelmann's astronomical career. As Schiebinger (1987) argues, such a shift of the observatory from domestic space to public space

[3] For a more detailed examination of this topic, see Larsen (2009), from which part of this discussion is summarized.

essentially eliminated access to observatories for women. As a result, when Gottfried died in 1710, Maria was heartily rebuked by the Council of the Berlin Academy of Science when she dared to apply for the position of assistant calendar maker, a position for which she was eminently qualified in skill and experience, but deemed unsuitable for simply because of her gender. When her son Christfried was named joint head observer of the Berlin Observatory in 1716, his mother tried to assume the role of his assistant, but she was soon reprimanded by the Academy for taking too public a role, and was warned to retire to her kitchen and care for her son's gastronomical rather than astronomical needs (Schiebinger 1987). In this way Maria Winkelmann Kirch was ultimately forced back into the domestic space and prevented from taking part in the aspect of science in which she had been trained and had demonstrated talent.

The case of Delvalle Lowry is, in a very real sense, a fortunate reversal of that of Kirch that deserves analysis (despite the fact that their lives were separated by a century). Lowry and her husband were technically in different fields (considering her to be a popularizer of geology and not primarily an artist or engraver). More importantly, she had already established herself as a professional at the time of her marriage (as she had published a successful book under her maiden name), and returned to her career after the death of her husband, rather than trying to establish herself under her married name based on credentials earned *during* the marriage. Whereas Kirch attempted to emerge from the domestic sphere into the masculine sphere of the professional observatory, Lowry and her husband can both be considered as having careers based *within* the domestic sphere, since his art gallery was part of their home.

Connections can also be drawn between female fossil hunters of the late eighteenth and early nineteenth century who operated within a domestic sphere (for example the Philpot sisters, and Mary Anning, who had been trained by her father) and Britain's most famous woman astronomer of the same timeframe, Caroline Lucretia Herschel (1750–1848). She gained acceptance both because her work was clearly rooted in the amateur tradition of observing celestial bodies for their own sake and only making accidental discoveries in the process, and because she clearly limited her role to that of assistant to her brother using their garden telescopes (Knapen 2010). Herschel was born in Hanover, Germany, and had followed her older brother William to England in order to pursue a career as a singer, but instead became William's assistant in telescope making and observing.

After discovering the planet Uranus in 1781, William became an international sensation, and gave up his own musical career to focus on astronomy, aided with a royal pension. Caroline assisted him at the telescope, at first by recording his observations and doing simple arithmetic calculations, and later by observing the heavens, searching for new star clusters and nebulae as well as comets. She made eight independent discoveries of comets, and in five cases was the first person to see the comet in question. After her first cometary discovery she became something of a celebrity, and despite her gender was granted the official title of Assistant Astronomer (with a small pension) by King George III (Ogilvie 1975). Caroline herself humbly claimed that she had done nothing "but what a well-trained puppy-dog would have done" (Clerke 1895:140) for her brother and referred to herself as "a mere tool which my brother had the trouble of sharpening" (Holden 1885: 182). Despite her celebrity, Caroline saw herself as simply occupying the limited and socially acceptable role of assistant to her famous brother, similar to how the women of the early 1800s who aided their famous geologist husbands were seen.

Another analogy to Delvalle Lowry's life is seen in the familial connection that shaped the career of Maria Mitchell (1818–1889) of Nantucket Island, widely regarded as the first American woman astronomer. The daughter of respected amateur astronomer William Mitchell and his wife Lydia Coleman, she was the descendent of the Quakers who founded Nantucket. Similar to the relationship between Delvalle and her parents, Maria studied astronomy in guild-fashion, as apprentice and assistant to her father, from whom she learned to make careful measurements and calculations of stellar motions and positions, skills desirable by both sailors and astronomers. As a result, Mitchell became quite skilled with telescopic observations (Albers 2001). In October 1847, this librarian was catapulted to world fame when she discovered a comet with her father's small refracting telescope, and was eventually awarded a gold medal from the King of Denmark for this feat. Although soon afterwards elected the first woman member of the American Academy of Arts and Sciences, full-time astronomical employment was still beyond her grasp. She was only able to procure part-time employment (at home), as a so-called *computer*, calculating the positions of (appropriately) Venus for the *American Ephemeris and Nautical Almanac*. She became the first astronomy professor and Director of the Observatory of Vassar College when it opened in 1865 and instituted an observing program for her students that contributed useful astronomical data to the scientific journals of her day.[4] Her female students were able to conduct nighttime observations partially because the observatory was literally attached to Mitchell's living quarters (thereby qualifying it as a domestic space) and because for many years her elderly father lived with her and served as a surrogate father and chaperone to the girls.

As her mother had done (and she likewise encouraged the readers of her books) Delvalle could easily study minerals within the domestic sphere of her family house using her parents' collection. However, her having been in the same field as her mother (and having so similar of a name) has led to a modern confusion between the two. For example, modern sources often attribute *Conversations on Mineralogy* to her parents (eg. Dolan 2000; Kuntz 2010). Such confusion is potentially fostered by the following comment by John Mawe in his preface to *A Descriptive Catalogue of Minerals*: "Information Relative to Mineralogy is given in the style of Lessons by Mrs. Lowry, with the advantage of an extensive and well arranged collection" (1816: ix) and exacerbated by Delvalle's young age at the time of her writing the book. Perhaps modern readers might be excused if they felt it unlikely that a woman so young could pen a science book in such detail.

---

[4] For more information on Maria Mitchell, see Wright (1949), Gormley (1995), and Bergland (2008).

While many women found their place in science largely through the encouragement and reputation of their family members, a somewhat reverse case can be found in Charlotte Hugonin Murchison (1788–1869) and her husband, geologist Sir Roderick Impey Murchison (1792–1871). She was the daughter of General Francis Hugonin and his wife Charlotte Edgar, from whom Charlotte learned botany. According to Roderick, the "old General of the 4th Dragoons, was a remarkably intelligent man, and a fair astronomer; her mother a most skilful florist and botanist" (Geikie 1875 v2: 332). Charlotte met her future husband, a military man who was over three years her junior, on the Isle of Wight in 1815, where he was visiting his mother. He was neither particularly well-educated nor well-read, but had attended Humphrey Davy's lectures in chemistry at the Royal Institution in 1812–3 (Geikie 1875 v1). In contrast Roderick wrote of Charlotte in his journal that she was "attractive, piquante, clever, [and] highly educated" (Geikie 1875 v1: 68). They were married the following summer and Roderick soon after left the army. The couple never had children but for a time Roderick's niece lived with them (Kölbl-Ebert 1997). As the property he had inherited from his family had failed to produce reasonable rent income and he had no solid plans for a second career (after briefly considering becoming a parson), he and Charlotte instead decided to travel to Italy for a year or two where their funds (mainly from her family) would sustain them longer.

In 1816 they traveled through France and across the Alps and into Italy, Charlotte sharing her knowledge of botany and geology with her husband, who had not yet showed much interest in science, along the way (Kölbl-Ebert 1997). After spending the winter in Genoa, they traveled to Rome and the accompanying countryside, where Charlotte caught and nearly died from malaria, an illness that complicated the remainder of her life. In Rome the Murchisons met Mary Somerville, with whom Charlotte became lifelong friends. Of the couple Somerville later wrote:

> Our great geologist, Sir Roderick Murchison, with his wife, were among the English residents at Rome. At that time he hardly knew one stone from another. He had been an officer in the Dragoons, an excellent horseman, and a keen fox-hunter. Lady Murchison,- an amiable and accomplished woman, with solid acquirements which few ladies at that time possessed—had taken to the study of geology; and soon after her husband began that career which has rendered him the first geologist of our country. (Somerville 1874: 122)

What Mary had not made clear here was the fact that it was due to Charlotte's influence that Roderick eventually took up the serious study of geology. When the couple returned to England, Roderick was bored, and for five years wasted his time and money on one of his only passions, field-sports, especially fox hunting. For her part, Charlotte became increasingly interested in mineralogy and built an herbarium, but her husband did not demonstrate significant interest in either (Geikie 1875 v1). In 1823 Roderick had the opportunity through a friend to hunt partridges with Sir Humphry Davy, and, in his words, "I perceived that a man might pursue philosophy without abandoning field-sports" (Geikie 1875 v1: 94).

Davy successfully convinced Roderick to utilize the observations of the Alps and Apennines that he had previously accomplished with his wife and come to London to take classes. While chemistry was admittedly neither his strong suit nor his passion, Roderick learned enough for field-geology. In the summer of 1825 Charlotte became his companion and assistant as he examined the local geology in depth, first as they walked their lands and neighborhood, and later on a nine week tour along the South Coast of England. Murchison's biographer writes that

> Driving, boating, walking, or scrambling, the enthusiastic pair signalized their first geological tour by a formidable amount of bodily toil. Mrs. Murchison specially devoted herself to the collection of fossils, and to sketching the more striking geological features of the coastline, while her husband would push on to make some long and laborious detour. (Geikie 1875 v1: 127)

During this expedition Charlotte met and became friends with Mary Anning, and collected fossils with her. This was the first of many expeditions Charlotte accompanied her husband on, collecting samples as well as making sketches. According to her obituary, during her husband's important early expeditions, Charlotte "was ever by his side, whether (with hammer and sketch-book in hand) on the southern and eastern coasts of England -the mighty Alps- Italy and Germany—[or] the mountains of her husband's native highlands" (*The Register* 1869: 297). In 1828, the couple was accompanied by Charles Lyell on an expedition across Europe to observe the stratigraphy and geological structures. Charlotte Murchison played an active role as assistant to the geologists, contributing in such varied ways as making sketches, collecting fossils, and even acting as French translator (Kölbl-Ebert 2007).

Roderick Murchison is best known for his 1830s study of ancient rocks in South Wales, where he identified a sequence of rocks that he dubbed the Silurian, after a local ethnographic group the Silures. His definitive work, *The Silurian System* (1839), solidified his reputation, with Murchison himself later noting that "in originating and completing the Silurian classification, from 1830 to 1838, she was very frequently at my side" (Geikie 1875 v2: 335). Murchison was not only a well-known geologist, but a founding member and for many years president of the Geographical Society. This necessitated the Murchison's frequently hosting foreign scientists and large social events in their home, which Charlotte accomplished with ease due to her own scientific background. It was written that "her intelligent mind and great knowledge was such that she could converse with philosophers—especially astronomers—on their own subjects, and set them at ease in a manner which only those who have felt the power of her conversation can duly appreciate" (*The Register* 1869: 297).

Charlotte spent the last few years of her life as an invalid, and died in 1869. She was then eulogized as

> no ordinary woman, and the world of science owes her a deep debt of gratitude, for if her ladyship (then Mrs. Murchison) had not—nearly half a century ago—weaned her husband's powerful mind from the ordinary occupations of a retired Peninsular Captain, and attracted his attention to the engaging pursuits of science, England might not have had occasion to be proud of the illustrious Baronet. (*The Register* 1869: 297)

Her husband agreed with this assessment of her influence "as the real source of all his scientific work and fame" (Geikie 1875 v2: 331).

In a letter to Roderick after Charlotte's death, Murchison's collaborator Adam Sedgwick spoke eloquently and sorrowfully about "one of the dearest of those friends whose society formed the best charms of my life" (Geikie 1875 v 2: 337). Murchison's biographer echoed the sentiment of her obituary, offering of Sedwick's genuine respect and affection for her

> She could have been no ordinary woman whose memory drew such an encomium from such a man. Her influence upon the career of her husband was not her only title to the grateful recollection of lovers of science. To the courteous bearing of a cultivated woman she added a brightness of conversation, an intelligence, and a range of knowledge which gave her a peculiar charm, and enabled her to please people of the most varied tastes and acquirements. (Geikie 1875 v2: 339)

In her role as mentor, tutor, assistant, and patient inspiration to her husband, Charlotte Murchison demonstrates not only the power of familial relationships in opening the doors to science, but the fact that these doors sometimes open both ways.

References

Albers H (2001) Maria Mitchell a life in journals and letters. College Avenue Press, Clinton Corners
Bakewell R (1819) An introduction to mineralogy. Longman, Hurst, Rees et al., London
Bergland RL (2008) Maria Mitchell and the sexing of science: an astronomer among the American romantics. Beacon Press, Boston
Brody J (1985) The pen is mightier than the test tube. New Scientist 105:56–58
Carlyle EI (1899) Weale, John. In: Lee S (ed) Dictionary of national biography, vol 60. Smith, Elder, and Co., London, p 89
Clerke AM (1895) The Herschels and modern astronomy. Cassell and Co., London
Davis HS (1898) Woman astronomers (400 AD–1750). Pop Ast 6(3):129–138
Dolan B (2000) The language of experiment in chemical textbooks: some examples from early nineteenth-century Britain. In: Lundgren A, Bensaude-Vincent B (eds) Communicating chemistry: textbooks and their audiences, 1789–1939. Science History Publications, Canton, pp 141–164
Fontani M, Costa M, Orna MV (2014) The lost elements: the periodic table's shadow side. Oxford UP, Oxford
Geikie A (1875) The life of Sir Roderick I. Murchison. Murray, London
Gormley B (1995) Maria Mitchell: the soul of an astronomer. Eerdmans Publishing Co, Grand Rapids
Hamilton J (2015) A strange business: art, culture, and commerce in nineteenth century London. Pegasus Books, New York
Henderson JP (1997) The life and economics of David Ricardo. Springer, New York
Holden ES (1885) The three Herschels. Cent Mag 39(2):182
Hunt W (1886) Boyd, Hugh Stuart. In: Stephen L (ed) Dictionary of national biography, vol 6. Macmillan, New York, pp 92–93
Knapen JH (2010) Scientific collaborations in astronomy between amateurs and professionals. In: Eversberg T, Knapen JH (eds) Stellar winds in interaction, pp 77–83. http://www.stsci.de/pdf/arrabida.pdf. Accessed 30 Mar 2017
Kölbl-Ebert M (1997) Charlotte Murchison (née Hugonin) 1788–1869. Earth Sci Hist 16(1):39–43
Kölbl-Ebert M (2007) The geological travels of Charles Lyell, Charlotte Murchison and Roderick Impey Murchison in France and northern Italy (1828). Geol Soc 287:109–117. doi:10.1144/SP287.9

Kuntz H (2010) Science culture in English-speaking Montreal, 1815–1842. Dissertation, Concordia University

Larsen K (2009) 'A woman's place is in the dome': gender and the astronomical observatory, 1670–1970. MP 2(5):104–124. http://academinist.org/wp-content/uploads/2009/10/Woman_Place_Larsen.pdf. Accessed 1 May 2017

Lowry D (1822) Conversations on mineralogy, 1st Amer. edn. Hunt, Philadelphia

Mawe J (1816) A descriptive catalogue of minerals, 2nd edn. Longman, Hurst, Rees et al., London

Monkhouse C (1899) Varley, John. In: Lee S (ed) Dictionary of national biography, vol 8. Macmillan, New York, pp 150–153

O'Donoghue FM (1893a) Lowry, Joseph Wilson. In: Lee S (ed) Dictionary of national biography, vol 34. Macmillan, New York, pp 212–213

O'Donoghue FM (1893b) Lowry, Wilson. In: Lee S (ed) Dictionary of national biography, vol 34. Macmillan, New York, pp 213–214

Ogilvie MB (1975) Caroline Herschel's contributions to astronomy. Ann Sci 32:155

Peacock R (1905) Modern British women painters. In: Sparrow WS (ed) Women painters of the world from the time of Caterina Vigri, 1413–1463, to Rosa Bonheur and the present day. Hodder and Stoughton, London, pp 65–72

Schiebinger L (1987) Maria Winkelmann at the Berlin Academy. Isis 78:174–200

Schiebinger L (1989) The mind has no sex? Harvard UP, Cambridge

Smith EC (1931) Some notable women of science. Nature 127:976–977

Somerville M (ed) (1874) Personal recollections, from early life to old age, of Mary Somerville. Murray, London

Story AT (1894) James Holmes and John Varley. Bentley and Son, London

The annual biography and obituary for the year 1825 v9 (1825). Longman, Hurst, Rees et al., London

The Atheneum (1824) Wilson Lowry. 2(5):192

The Literary Gazette (1823) Advertisements. Conversations on mineralogy. No. 323:208

The Literary Panorama (1814) Astronomy. 15:549

The Quarterly Journal (1823) Conversations on mineralogy, with plates, engraved by Mr. and Miss Lowry, from original drawings. 14:154–160

The Register, and Magazine of Biography (1869) Lady Murchison. 1:297–298

The Spectator (1859) Deaths. http://archive.spectator.co.uk/article/31st-december-1859/20/births. Accessed 20 Mar 2017

Varley D (1846) The engineer's manual of mineralogy and geology. Weale, London

Varley D (1859) Rudimentary treatise on mineralogy, 4th edn. Weale, London

Wilson LG (2013) John Jeremiah Bigsby, MD: British Army physician and pioneer North American geologist. In: Duffin CJ, Moody RTJ, Gardner-Thorpe C (eds) A history of geology and medicine. Geological Society of London, London, pp 375–394

Wright H (1949) Sweeper in the sky: the life of Maria Mitchell. Macmillan, New York

Chapter 7
Delia Woodruff Godding (1812–61): Poetry and Perseverance

7.1 Biographical Overview

The details of Delia Woodruff Godding's fascinating and bittersweet life have remained unstudied until now. In fact, very little information on Delia has been located; prior to the author's interest in Godding's geological writing, published information on the author was limited to what was found on the cover of her 1847 children's book published in Hartford, CT, *First Lessons in Geology* by a "Miss D.W. Godding" (eg. Ogilvie and Harvey 2000). Published here for the first time is a comprehensive biographical sketch of this author, as well as an analysis of her popularized work in geology.

In an April 1, 1855 letter to his older sister, William A. Godding (1816–1897) urged her to "Write often Delia as I do to you for there is but a few of us left and we should speak often one to the other."[1] These prescient words were born from a man whose immediate family had already suffered considerable loss. Two weeks before the writing of this letter, William's older brother, Carlos, had buried his second wife, Thankful P. Harvey (1814–1855), after less than five years of marriage. Her sister, Sabrina Harvey (c. 1811–1848), Carlos' first wife, had died seven years prior, and his only child, Wilbur B. Godding (c. 1839–1913) had now lost both his mother and stepmother/aunt by the age of sixteen. By the time William himself passed away at age 81 he had already buried his own wife, Sarah Jane Kimball (1835–1870), who succumbed to Typhoid fever, and two children—Martha Jane (1860–1862) as an infant and Delia Godding Fairbanks (1855–1885)[2] to consumption—as well as outlived his siblings Delia and Carlos. Another older brother (confusingly also named William A. [1807–1810]) had died before he was born. His family's deep abiding faith and his own "perfect picture of health" (inherited from his parents, although he

[1] The three letters referenced in this section are held by the Connecticut Historical Society and comprise the entire collection labeled *Delia Godding correspondence, 1847–1855* (Ms 101610).

[2] The Godding family's preference for the names Delia and William has proven to be a challenge when reconstructing their history.

K. Larsen, *The Women Who Popularized Geology in the 19th Century*,
DOI 10.1007/978-3-319-64952-8_7

felt that he should "thank my creator for it") personally served him well, but at the time of his death his closest living family was his only nephew, Wilbur.[3]

Delia Woodruff Godding was born in mid-June (possibly June 16), 1812, the only daughter of farmer William Godding (1773–1850) and his wife Martha Gridley (1779–1850).[4] William had been born in Attleboro, MA and Martha in Connecticut. They were married in 1803, and their farm on the east slopes of Burke Mountain in northeast Vermont was one of only a handful then established in the area. The formal village of East Burke itself was not established until the 1820s (Godding 1867). William and Martha's first son, Carlos, tended the farm alongside his father, while their youngest, William A., worked in business, including the lumber industry. According to Burke historian Jason H. Sanderson, Delia Godding attended the White School in Burke, the largest of the schools in town, and was known among the good people of Burke as a "woman of fine Christian character" (Personal communication. June 8, 2014). Little else is known of her formal education. Of all the Godding siblings, only Delia left their hometown, to become a teacher. While a biographical note published in the *Vermont Historical Gazeteer* states that she "educated herself for a successful teacher without any pecuniary aid" and afterwards taught in several Vermont towns, Hartford (Connecticut), St. Louis (Missouri), and St. Anthony (part of Minneapolis, Minnesota), specific historical records have only been found for the second and fourth of these (Hemenway 1867: 312). It is known that she received a diploma for "fine Pencil Drawing" from the Hartford, CT, County Agricultural Society in 1852 (*Hartford Daily Courant* 1852: 2).

It is unclear precisely when Godding left Vermont for Connecticut, where she had several cousins, including a married Hartford shopkeeper named Mary Anne with whom she was close. A letter from Delia's mother (from the farm in East Burke) dated January 4th to Delia in Hartford does not have a clearly marked year, although the numbers could be read as1844. Martha Godding's trepidations at her daughter's distance (apparently written shortly after her daughter had moved to the "beautiful city") is palpable in her letter. She writes in part

> Although we are separated in person yet you are never absent from my thoughts and it is my continual practice to recommend you to the care of that being whose eyes can see all his creations and to whom the secrets of all hearts are open. I am a mother and my concern for your happiness is inseparably connected with my own. After you left home and friends I was very lonesome.... If you are discontented come home. I should be glad to receive you.

Tentatively dating the letter to 1844 is appropriate, because in a December 21, 1844 newspaper ad advertising its school's third term, St. John's Episcopal Church, 97 Main St., Hartford[5] lists Miss D.W. Godding as its teacher. Young children ages five

[3] Unless otherwise specified, all of the information in this section is derived from digitized marriage, birth, death, gravestone, and census records located through www.ancestry.com

[4] Martha died on May 13, 1850 while her husband died a month later on June 16.

[5] The parish moved to West Hartford in 1907 and the Hartford property was sold to the neighboring Wadsworth Atheneum for its expansion. A summary of the church's history is found at http://www.sjparish.net/About_us/Church_history/

to seven received instruction in "the usual elementary branches" while older children and "young ladies are received and instructed in the higher English branches; also in Drawing and French if desired" (*Hartford Daily Courant* 1844: 3). The only other definitive record of Godding having taught in Hartford was in *Greer's Hartford City Directory for 1857–8* (Greer 1857) where she is listed as St. John's assistant for the Sabbath school.[6]

Delia Godding wrote her *First Lessons in Geology* during her time in Hartford, with the preface dated "Hartford, November 1846" (Godding 1847: vi) and the book officially issued by Hartford publisher Henry S. Parsons in 1847. A letter written by Delia (then in East Burke) to her cousin Mary Anne in Hartford dated December 27, 1847 suggests that Delia had spent some period of time back home in Vermont after the publication of her book. Delia notes that her cousin must be very busy at her shop during the Christmas and New Year season, and playfully inquires if Mary Anne has "got any of my scribbling to bind," presumably a reference to her book. Delia also sends words of solace to her beloved cousin on the loss of a child, apparently not her first such loss. Delia's profound faith is reflected in her words of comfort:

> The hand of God seems to be upon you in oft repeated affiliations. I trust, yea I believe, that it is for some wise, and kindly purpose, perhaps to draw you nearer to himself and help you to place a due estimate on the things of time. I hope you all see and feel it to be so. You have the unspeakable comfort of knowing that the little ones are forever safer and this thought should soften your grief and fill your hearts with holy, joyful resignation to the will of God. (Emphasis original)

Delia apologizes for her tardiness in writing, explaining that she has spent much time making dresses for friends and family, complaining that

> People think that because I am from the city that I have all fashions and can make anything to order. I [did?] work for a few who were needy to accommodate and so I got myself into business very quick. I was soon called on to cut wedding dresses and make wedding head-dresses and c. In the midst of it I abruptly took my leave and went to Danville [to visit extended family].

As reflected in the 1855 letter from her brother, the health of the Godding family was ever on their minds, and constantly a source of worry. Delia's letter to Mary Anne explains that at that time a young relative, Wesley, was sick from smallpox that he had caught while teaching at Newbury Seminary, while another relative, Haynes, had come down with rheumatic fever. Her sister-in-law, Sabrina, was recovering from some unstated ailment (and unknown to the family at the time, would die the next year) and Sabrina's sister, Thankful, was herself undergoing treatment for some condition.

It is not known when Delia returned to Hartford and St. John's Episcopal Church to teach Sabbath School, but the letter from her brother was sent to her in Hartford

---

[6] Bill Uricchio, parish historian for St. John's Church, was unable to provide further information, explaining that the parish's historical records are quite incomplete (private communication March 26, 2017). His assistance in locating some of the newspaper references cited in this chapter was of immeasurable benefit.

in April 1855 after the death of Carlos' second wife, and insinuates that she had been gone from Vermont for some time. William notes of Carlos that he "is very lonely now more so than he ever was before" and expresses his wish that Delia could live with their older brother. Instead, William observes, a "very Good woman" named Mrs. Blake had been living with Carlos and will hopefully remain "through the summer if her daughter does not send for her." Apparently her daughter did not, because Carlos married a Mrs. Julia A. Blake on December 9, 1855 and his son Wilbur married a young woman named Addie Blake (b. 1847) in July 1868. The third Mrs. Carlos Godding presumably outlived her husband (who himself died at age 71 of dropsy of the chest on May 13, 1876, the anniversary of his mother's death) as she is not listed on his gravestone alongside the Harvey sisters.[7]

In the last years of her life, Delia traveled to St. Anthony, a section of Minneapolis, and was the principal of a private Episcopalian school for girls called the Sigourney Boarding School. It was one of many female seminaries that were established in Minnesota during this time. For example, a Presbyterian boarding school and female seminary was founded in St. Paul in 1856 (Fawcett 1933). The school first opened for the 1860–1 school year (*Collections of the Minnesota Historical Society* 1880: 52).[8] She hired an experienced teacher, Miss Elizabeth Brooks (1838–?), a Canadian who had previously taught in Virginia, Connecticut, and Maine, to teach music, French and advanced English during the second half of that first school year. (Greenlee and Greenlee 1908). In late June 1861, Delia was visited by an old acquaintance from Vermont, Jason Adams Niles (1814–1894), whom she had first met in 1837 (the year he graduated from college in Burlington) and again in 1856 (Niles 1864).[9] She returned to Vermont near the end of the summer to visit her family, and died on September 15, 1861 at the age of 49. The official cause of death was listed as Billious and Typhoid Fever, a common diagnosis at the time. She was buried in Woodmont Cemetery in East Burke; her headstone notes that it was "Erected by the Sunday school of St. John's, Hartford Ct. to their Teacher and dear friend," suggesting that she had taught there for many years and as recently as just before moving to Minnesota.[10] Elizabeth Brooks initially took over Godding's role as principal of the Minnesota school "until the Indian outbreak late the following Spring. Her parents, fearing for her safety, prevailed upon her to return East" (Greenlee and Greenlee 1908: 598).

---

[7] The gravestone for Carlos, Sabrina, and Thankful Godding can be viewed at https://findagrave.com/cgi-bin/fg.cgi/http%25252522//fg.cgi?page=gr&GRid=32797527

[8] Many thanks to Bill Uricchio, parish historian for St. John's Church, for bringing this item to my attention.

[9] After his college graduation, Niles had taught school in Tennessee for several years before becoming a lawyer. He eventually settled in Mississippi where he was a judge and served as Congressman from 1872–5 (*The Kosciusko* 1894).

[10] Delia Godding's grave was recently visited by Bruce Beford and the inscription on the headstone was covered in lichen, rendering it impossible to clearly photograph. A legible rendition is found at https://www.findagrave.com/cgi-bin/fg.cgi?page=gr&GRid=22631049

Godding's apparently short-lived school took its name from Lydia Howard Huntley Sigourney (1791–1865), a Connecticut native and famous author nicknamed the Sweet Singer of Hartford. She was a teacher in her youth, including opening a girls school in her native Norwich with her friend Ann Maria Hyde, afterwards teaching neighborhood children (including African-American children) until she opened a school in Hartford in 1814. She ran this school until her marriage to wealthy Hartford businessman and widower Charles Sigourney in 1819 (Collin 1902). She was widowed in 1854. Sigourney was a popular and influential writer and poet, and throughout the latter half of nineteenth century women's literary societies and college clubs were named in her honor. She was well-known for her support of women's education, and was a benefactor of Mary Lyon's Mount Holyoke Female Seminary. Her book *The Weeping Willow* was published by Henry S. Parsons as well, about same time as Godding's *First Lessons in Geology*, so there are several reasons why they could have known each other. Sigourney was also known for her elegies, and she wrote one for Godding that was included in *The Man of Uz and Other Poems* (1862). The preface statement and opening verse read as follows:

> A faithful Teacher of the young from early years, and recently the Principal of a Female Seminary and Boarding School at St. Anthony, Minnesota, died suddenly of an attack of fever, while on a visit at her paternal home in Vermont, September, 15th, 1861.
>
> THINE earnest life is over, sainted Friend!
> And hush'd the teaching voice that gladly pour'd
> Knowledge and goodness o'er the plastic mind.
> —Full many a pupil of thy varied lore
> Amid thine own New-England's elm-crowned vales
> Holds thee in tenderness of grateful thought,
> And far away in the broad-featured west
> Where the strong Sire of waters robes in green
> The shores of Minnesota, comes a wail
> From youthful bands expecting thy return,
> To guide them, as the shepherd leads the lamb. (Sigourney 1862: 222)

The drama surrounding Godding's life continued after her death. Unbeknownst to her family, she had purchased land in St. Anthony's, presumably with the intention of finally settling down in one place. After her death, the taxes were not paid on her property, and on August 20, 1887, Edgar P. Inglee filed a tax lien with the District Court of Hennepin County against Godding. The court called for a summons to be posted in the *Saturday Evening Spectator* for six weeks, and since no one came forward to pay the taxes, on October 29, 1887 ownership of the lot was given to Inglee. On August 5, 1892, William A. Godding and his nephew Wilbur filed an affidavit stating that Delia W. Godding had died out of state and that they, her only heirs and also out of state residents, had had no knowledge of, or consultation in, the matter. Their motion to set aside Inglee's award was denied and they appealed to the Minnesota Supreme Court. Their appeal was ultimately denied (Willson 1894).

7.2 First Lessons in Geology (1847)

Delia W. Godding's *First Lessons in Geology* is unique among the geological works described in this volume, in that it is among the briefest (less than 150 pages) and the only one utilizing a catechism format of simple questions and answers. As noted in Sect. 2.2, this is the least common design utilized in the familiar format genre. However, Godding does successfully adapt the potentially stifling and staccato plan, wherever possible having questions connect back to answers in a way that maintains some modicum of flow. In the case of this volume, the expert maternal figure who imparts the knowledge is Godding herself, who refers to herself in the first person throughout the work. The other side of the conversation is taken up by the reader, rather than fictional children.

In her preface Godding spells out two reasons for producing the text. Her first is to "fill a void, which, as teacher she has in vain endeavored to supply from other sources: there being many excellent works in this science, but none sufficiently simple for the ready comprehension of children" (Godding 1847: v). The second clearly aligns with the faith that sustained not only her but her entire family (as evidenced from her family letters): "to awaken in young minds an early interest in the science of Geology, which is daily becoming more and more a witness to the truth of God, and an illustration of his power and majesty" (Godding 1847). Godding invokes her own experience teaching children in further explaining that she has found "that no subject is more pleasing, and interesting to them, than explanations and illustrations of this science when fitly presented, and this fact induced her to prepare this work for their pleasure and benefit" (Godding 1847).

As noted, the familiar format embraces the concept of adapting or simplifying the subject in order to make it more suitable for young minds to grasp. Godding's book is no exception. She explains that her intention is "to open the door, leaving to others, to explain more minutely the many yet unexplored beauties of the inner temple where nature is still working out the laws, given to her when God said, 'Let the dry land appear, and it was so'" (Godding 1847: v–vi). Her particular brand of adaptation is described as a plan to present only "known and acknowledged facts, unmingled with theory" in the hopes of being able to "awaken the slumbering genius of some young minds, and lead them on to make still deeper researches" (Godding 1847: vi).

The welcoming conversational style of her catechism is set up from the start. Chapter 1 begins

> Come, my pupils, let me tell you a pleasant story about the *wonders of the world,* or the little globe which we inhabit. If you will give me your attention while we talk familiarly, I think you will be amused and instructed. I hope also, you will be made wiser and happier. You have probably all heard, and read, the word *Geology* many times, and perhaps you have as often wondered what such an odd word can mean; and now you are all ready to ask,
>
> What *is* Geology?
>
> It is a science which treats of the earth's structure, and of the substances which compose it. (Godding 1847: 9–10; emphasis original)

Note the personal style, in which the teacher is inviting her students to take part in a conversation about geology. She speaks of it as a *story* that will both amuse and instruct, and begins by referring back to a word that at once seems familiar and strange: *Geology*.

Similarly at the start of Chap. 8 on geological strata and the center of the earth, she reminds her students of a previous lesson: "You will remember that at the close of Chapter III, I promised to tell you more about the internal parts of the earth, and the opinion of geologists in reference to its centre. Now if you study this chapter attentively you will learn my story..." (Godding 1847: 39). The use of *promise* and *story* again add to the particularly personal rather than informal nature of this conversation. Another example is found at the start of Chap. 10 on fossils. She begins "Well my young friends, I have been telling you something about minerals, and you have passed over the hard names very well. Now we will talk about something very wonderful and amusing" (Godding 1847: 49). Again, the use of the personal term of endearment *young friends* and the entreaty *we will talk* allow the reader to imagine that she or he is there in the classroom with Miss Godding as their personal teacher.

As Godding's work is written at an elementary level and very short, there is no grand list of references at the beginning. Nonetheless there is the usual list of well-known authorities cited within its pages, including Edward Hitchcock, Gideon Algernon Mantell, Benjamin Silliman, Charles Lyell, William Buckland, Georges Cuvier and others, whom she has clearly read. She often recommends Mantell's popular works by name, and in particular calls *The Medals of Creation* "an excellent work for the young" (Godding 1847: 67). She also defers to these experts and tells her readers that they should do the same. For example, when recounting the different strata, she references Mantell, who "in his *Wonders of Geology*, has given an interesting subdivision of these groups" (Godding 1847: 41). An anonymous avatar of the reader later inquires "How far down have the series of rocks been examined?" to which Godding remarks "About fifteen or twenty miles. See Dr. Mantel's [sic] Medals of Creation" (Godding 1847: 44). Another anonymous questioner asks "Of what does Dr. Mantel [sic] say our globe is composed?," the answer being "Of minerals and fossils. You shall hear of fossils in another chapter" (Godding 1847: 45). The discussion of fossils includes illustrations of the labyrinthodon (captioned *labyrinthidon*) and footprints copied from Edward Hitchcock's *Elementary Geology*. However, as previously noted (Sect. 5.3), it was common for American authors to engage in such borrowing, and Hitchcock himself borrowed his sketch from one by paleontologist Richard Owen, although with attribution (Hitchcock 1845).

Given the literary style of Godding's work, there are few natural places to make connections to everyday life. One example is the question "How may geology benefit the farmer or agriculturist?" The given answer is that it "will teach him how to temper the soil to different kinds of grain, to make it rich and productive" (Godding 1847: 95). However, another such instance is easily missed by modern readers. In the aforementioned discussion of fossil footprints, the anonymous student inquires

> What kind of creature could have made such tracks as those?
>
> Professor Silliman, Mr. Owen, and some others, think they might have been the tracks of great frogs, whose hind feet were much larger than their fore feet.

What was probably their size?
They must have been about the size of a hog or calf; some say they were as large as an ox.
That would be laughable indeed! (Godding 1847: 66)

The reference here is to a common fable, *The Ambitious Frog*, that tried to puff itself up to be the size of an ox.[11] Edward Hitchcock uses a similar reference to the same fable in his description of the creature in *The Religion of Geology and Its Connected Sciences* (1851):

> Still more unequal in size, as compared with living batrachians, was the labyrinthidon[sic], once common in England and Germany, if, indeed, the tracks on sandstone were made by that animal. It was, in fact, a frog as large as an ox, and perhaps as large as an elephant. Think of such animals swarming in our morasses at the present day! (Hitchcock 1851: 228)

This discussion of fossils also affords Godding an opportunity for a lesson in morality, when she asks

> Do you think you would like to walk out often, if they were abundant in these days?
> I think not, especially if they could jump far; yet if they were as harmless as our frogs, I think I would not fear them much.
> That is a right conclusion. Children, or older persons, should never fear animals who do not make war upon mankind, without just provocation. (Godding 1847: 66–7)

But it is a deep Christian religiosity that forms not only the backbone of the moral lessons, but the book itself. In keeping with the steadfast faith evidenced in her family letters, Godding takes every opportunity to remind her students of the wisdom, power, and benevolence of their Creator, and how these traits are reflected in the geological record. In the first chapter she notes of the continents, oceans and other formations that "Geology tells *how* and of *what* they are made" (Godding 1847: 10). The student response is appropriate in its piety: "How they were made! why, I thought God made *all* things. Did he not?" Godding is quick to put her students' minds at ease: "Yes, he is the great Author of every particle, however minute at the first; but he made certain laws by which he united every particle to its mate, and brought every substance to its present form." Godding then makes clear her opinion as to the proper interpretation of the Scriptures in connection to geology when the student asks "Does not the bible tell how God made the world?" (Godding 1847: 10). Godding affirms that

> Yes: a short account is given in the first chapter of Genesis; but geology gives a more extended and particular view of the subject….. The bible was given for our spiritual benefit, and as such is a perfect book, containing all things necessary to our salvation; but it was not intended for a book of science, and therefore does not treat at large upon scientific matters. (Godding 1847: 11)

Therefore when the student inquires "Does Geology agree with the bible history of the creation?" Godding confidently answers

> It presents nothing contrary, and nothing that cannot be easily reconciled with it…. it exalts the mind, and gives us more generous views of that infinite Being who sustains us all. It leads us to wonder at his power, and to adore his wisdom in all his works.

[11] See, for example, Parley (1857).

> Well then, I shall take an interest in what you tell me, if that be truth.
> Such is the *truth*, my dear pupils; and at the close of this little work, I think you will agree with me, that in every line Divine Wisdom shines brighter and receives and reflects glory from the smallest atom of creation. (Godding 1847: 11–12)

Note the use of the word *reconciled* to describe this relationship, rather than *proven*, echoing the late 1830s writings of Edward Hitchcock (Sect. 1.3). Again, we see here the use of a term of endearment, *dear pupils*, reminiscent of the language of Margaret Bryan, and the encouragement that the pupils will come to agree with her of their own volition at the end of the lessons.

While the students will arrive at this conclusion of their own free will, it is not without the careful direction of their teacher. Godding takes every opportunity to point out precisely where she believes Scripture and geology closely agree. For example, in discussing the formation of the earth, she notes that originally it had no regular form, but was instead a "confused mass of shapeless matter" (1847: 15). She then reminds the students that the Bible similarly calls the earth "without *form* and *void*" (Godding 1847: 16). A similar connection is found in a discussion of fossils, which suggests "that plants, trees and animals, were much larger, more coarsely formed, and longer lived at the beginning, than they are now. So different authors have said, and Bible history seems to confirm this idea" (1847: 131). If these examples were not sufficiently convincing to the young student, the last full chapter of content is termed "Agreement of Geology with the Sacred Scriptures." It is here that Godding's book most clearly reads like a Christian catechism:

> What important fact should we remember?
> That the works of nature declare as plainly that there is a God as Revelation, and they teach us as plainly his goodness and truth....
> What is another fact that we ought to learn?
> That nature's works and God's word agree in their testimony. (Godding 1847: 128)

We hear an echo of William Paley when Godding explains to her students "When we look at the works of nature we see God's goodness in 'tempering the winds to the shorn lamb,' and providing for the wants of all, in adapting the organs of the body to the circumstances in which they are placed" (1847: 129). When pressed for further examples, she includes the wings of the bee, the strength of the ant, and the hen defending chicks with her wings.

But while devoutly religious, Godding makes it clear that she is not a literalist in terms of the Biblical chronology. As in the case of Hitchcock, she explains to her students that there is no disagreement between geology and the Bible concerning the six days of creation, because "truth never contradicts the word of God" (1847: 132). She offers as one potential explanation that

> In the first chapter of Genesis we learn that the sun and moon... were not created till the fourth day, and as it is said they were put for signs and for seasons, and for days and years, time before that period could not have been measured as ours is now, and may have been measured by indefinite periods, that is, many years for one day, as Geology supposes, without contradicting the word of God. (Godding 1847: 132)

She then offers a "more satisfactory" explanation: "In the beginning God created the world. The term, in the beginning, means no fixed time, and may reach back a great number of ages before the present system of things, described as the six days work of creation" (Godding 1847: 133). Godding not only refers back to an older generation who once held that there was a conflict between geology and theology, but to the central concept of Hitchcock's philosophy when she notes

> every one receives it as truth now: so when prejudice dies away, and the young become better instructed in Geology, none will be startled at its truth…. Upon farther examination we should probably find, that there is no real disagreement between Geology and Bible history; and that in studying this science we are almost imperceptibly led to set a greater value upon its sacred instructions. (134)

In keeping with her familiarity with both the local geology of New England and its famous geologists, Godding references both in her work as they would include the geological formations, fossils, and personalities that her students (and by extension her readers) would have had the greatest familiarity with. As an example, she refers to Benjamin Silliman and Edward Hitchcock numerous times in her text. When discussing veins of trap rock, she uses the example of Salem, Massachusetts, where Hitchcock had been able to identify eleven different eruption events (1847: 36). Chapter 18 *Review of Rock Formations* begins with an explanation that she is basing her discussion, in part, on "the views of a gentlemen, who knows a great deal about geology, and does much to promote the science." A footnote identifies that gentlemen as Silliman (1847: 97). When discussing fossils, she includes (among others) fish found in the limestone on Mount Lebanon by Silliman and footprints found in the Connecticut Valley by Hitchcock, opining of the latter "This gentleman is constantly presenting something new and interesting in this science" (Godding 1847: 120). When discussing fossil footprints that resembled those of birds, she also makes another local reference, explaining that they have been found in "the new red sandstone in Scotland, in forest marble in England, and in red sandstone in the Connecticut valley" (Godding 1847: 64).

But the most curious local example involves a discussion of the "great American Mastodon," of which a complete skeleton had been found in August 1845 in Newburgh, New York. Godding includes a full page picture (Fig. 7.1), "drawn from the real skeleton, by the author of this little work, Hartford, Nov. 26, 1845" (1847: 72). Now called the Warren Mastodon, this skeleton has a permanent home on display in the American Museum of Natural History in New York City. Godding's book offers some rather interesting ideas in regards to this "wonderful animal":

> Some thought he was antediluvian….Others believed that he died only a few centuries ago, while some thought him the Behemoth of Job. But his history is veiled in obscurity. We only know that such a noble animal once walked the earth, and that its massive bones, after many centuries had passed away, have come forth at last from their silent resting place, to speak to us of the wondrous power and wisdom of the Creator. (Godding 1847: 74)

Ralph O'Connor (2013) points out that a similar, albeit highly romanticized, connection is made between fossil creatures and Biblical monsters in James Mellor Brown's *Reflections on Geology* (1838). In terms of her opinions concerning the

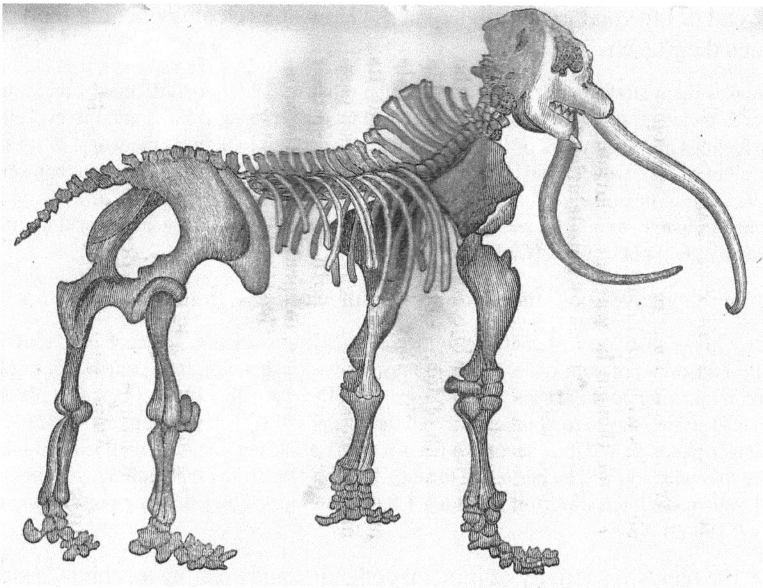

Fig. 7.1 Sketch of the Warren Mastodon by Delia W. Godding. Reproduced from Delia W. Godding (1847) *First Lessons in Geology*

Noachian flood and evidence in the geological record, Godding takes a confident approach, similar to that of Jane Kilby Welsh:

> What great event in the world's history has been noticed by geologists?
> The deluge, or Noah's flood.
> How many years after the creation did this occur?
> One thousand six hundred and fifty-six years. (Godding 1847: 122)

There is one very explicit hands-on activity described in this work. In order to model the oblate shape of the spinning earth, the student is directed to push a stick through a lump of clay and then rapidly whirl the stick between the hands until the clay slightly flattens at the top and bottom (see Fig. 2.2). Other references to hands-on activities and direct observations are few, and connected with her religious ideals. For example, after a discussion of veins in rocks (including dykes), she interrupts the conversation:

> Now my young geologists, I have told you a long story about veins, and I hope wherever you walk or journey, that you will take pleasure in searching for these veins in rocks, and ask some kind friend to explain them more perfectly to you.
> What should we carefully study?
> The glorious works of nature.... in them God teaches us many valuable lessons, and they are the best of school books....
> By thus doing, what may we learn to adore?
> The infinite wisdom of the great Author. (Godding 1847: 37–8)

At the end of the volume, the student asks "How does Geology benefit the young?" to which their teacher answers

> It makes them students of nature; elevates the mind to noble pursuits; makes them thirst after knowledge; gives them enlarged views of God's power, goodness and mercy; refines the feelings and sentiments of the heart, and leads them to look upon the world as it really is, full of beauty, wonder and life, yet full of change, decay and death; and upon themselves as its fleeting inhabitants, passing away—naturally teaching them to look away from this scene of change, to a more enduring inheritance of beauty and glory, in a world not measured by the flight of time. (Godding 1847: 136–7)

Godding then leaves her students with a homework assignment and her

> ardent hopes that you will continue to love and study this science. Let each one resolve to collect a cabinet of your own. This you may easily do when you take your walks of pleasure, by picking up the stones which lie scattered in your path.... It will give you pleasure in[sic] a stormy day to look them over, tell their names, their compositions and their uses to younger children, and thus become a teacher yourself. When you have well learned what I have told you, you will be permitted to study larger works from other authors, and they will tell you many interesting things which I have now been obliged to suppress. (Godding 1847: 138–9)

Again, the hands-on activity of mineral collecting and creating a cabinet is stressed, a common recommendation in these works.

As is the case with previously described books, reviewers across the decades have not been universally kind to Godding's book. A reviewer for the *Church Review and Ecclesiastical Register* expressed doubt that this "neat and attractive looking little volume" would be useful to its intended audience, but not

> from any hesitancy in respect to the skill of the authoress. On the contrary, her book exhibits familiarity with the subject, and much ingenuity in presenting it to the minds of children. But we doubt whether Geology itself is yet defined with sufficient clearness, in respect to some of its first truths, and many of its proofs, to allow such a simplification; and we also doubt whether children are capable of grasping the subject with any clearness. A work like this, though enlarged, prepared for high schools and academies, might be of service. (*Church Review* 1848: 288)

A particularly unkind review penned by Katharine Burrill in 1904 specifically panned the book's cover and frontispiece artwork (Fig. 7.2), complaining

> Anything more confused I never saw. For in one picture are displayed a volcano in eruption, a labyrinthodon, a dinotherium, birds, and gentlemen with top-hats. The whole is crowned with rocks, which are inscribed simply 'strata,' while the topmost peak is occupied by a small boy with a very large flag on which is written Persevere. I presume it means persevere in the study of Geology; but then why birds and labyrinthodons? They seem more Zoological. Still, the Persevere flag is not a bad one to carry whether you are learning lessons or doing your part in the world's work. (Burrill 1904: 9)

Given what has now been uncovered about Delia Woodruff Godding's life, it is more than likely that "Persevere" was her own personal mantra, and the optimistic illustration (note that the child is pictured as standing, triumphantly, on the top of the difficult mountain) was her personal way of sharing that difficult life lesson with her beloved students.

Fig. 7.2 Frontispiece. Reproduced from Delia W. Godding (1847) *First Lessons in Geology*

7.3 Poetry and Geology

Delia Godding was not only eulogized in verse, but composed a few of her own. *First Lessons in Geology* ends with an original poem entitled "Review in Rhyme, or The Seven Wonders of Geology." This seven verse poem references the birth of the earth, depth of primary rocks, forces that lift mountains, how fossils form, creation of coral reefs, subterranean molten rock, and finally the death of earth. Each verse opens with the refrain "I wonder." For example, the final verse asks

I wonder what hour,
By Omnipotent Power,
Creation's vast wheel shall be stayed;
And the internal fire,
Bursting forth in its ire,
Earth's funeral pile shall be made. (Godding 1847: 141)

The poem was reproduced several years after her death in the *Vermont Historical Gazetteer* (Hemenway 1867). Such an epilogue is not unexpected, for just as John Rennie argued that geology is romantic in his *Conversations on Geology*, so, too, have poets, especially in the Romantic Movement, incorporated geological metaphors and examples in their craft, especially as signifiers for the passage and effects of time (Heringman 2004). On the other hand, nineteenth century geology texts, such as those examined thus far, often included poetry quotations, especially of a religious nature, not only at the opening and closing of chapters but also interspersed within the body of the text. In his survey of 78 geology texts published between 1790–1860 (including Maria Hack's *Geological Sketches* and Mary Roberts' *Progress of Creation*), John Robert Higgins (2011) found that seventeenth and eighteenth century poetry was quoted most often, with Milton's *Paradise Lost* being the most commonly cited work, even outperforming the Bible itself.

The borrowing of geological examples by poets did not go unnoticed by the science writers of the day. John Aikin (co-author of *Evenings at Home*) argued that natural history "is not only a source of agreeable and innocent amusement, but conduces to humanize and enlarge the mind, and in various ways to promote the happiness of mankind, has been sufficiently proved by the observations of many ingenious writers. But its application to the improvement of poetry, has not, I believe, been the subject of particular discussion" (1777: iv). However, he openly criticized the poetic community of his time, noting a real lack in the diversity of the poetic imagery employed, which he attributed to the average poet's "scarcity of original observations of nature" (Aikin 1777: 9). In addition, this lack of experience with the natural world caused poetic descriptions to be

> faint, obscure, and ill characterized; the properties of things mistaken, and incongruous parts employed in the composition of the same picture. This is owing to a too cursory and general survey of objects, without exploring their minuter distinctions and mutual relations; and is only to be rectified by accurate and attentive observation, conducted upon somewhat of a scientific plan. (Aikin 1777: 10)

On the other hand, some gentlemen (and gentlewomen) of science wrote poetry that was dizzying in the breadth of its references concerning not only the natural world, but occasionally the literary one as well. The most famous example is Erasmus Darwin's *The Botanic Garden* (1791).

While modern literary scholars have argued that Darwin's poetry played a role in influencing the Romantic Movement, and historians of science find a treasure house of riches buried within his numerous appendices and annotations, critics of his time were far less generous with the poet (Primer 1964). *The Botanic Garden* (composed in two parts, *The Economy of Vegetation* and *The Loves of the Plants*) is not only (in) famous for its heavy-handed (and sometimes obscure) mythological references, but its copious footnotes explaining both the mythological and scientific references (at times taking up more room on the page than the poetry itself). The litany of scientific topics discussed in *The Botanic Garden* is legion, including such non-botanical concepts as electricity and magnetism, comets and meteors, the formation of coal and iron ores, geysers, and the surface of the moon.[12] The work itself advertised its general design to be "to inlist [sic] Imagination under the banner of Science; and to lead her votaries from the looser analogies, which dress out the imagery of poetry, to the stricter ones, which form the ratiocination of philosophy" (Darwin 1798: v). Consider the following excerpt:

---

[12] For a detailed analysis of *The Botanic Garden*, see Emery (1941).

Gnomes! how you shriek'd, when through the troubled air
Roar'd the fierce din of elemental war;
When rose the continents, and sunk the main,
And Earth's huge sphere, exploding, burst in twain.
Gnomes! how you gazed, when from her wounded side,
Where now the South-Sea heaves its waste of tide,
Rose on swift wheels the Moon's refulgent car,
Circling the solar orb, a sister (Darwin 1798: 41)

The reference here is to a hypothesis (later much expanded by his great-grandson, George H. Darwin) that the moon was formed when it split from the earth, creating the Pacific Ocean basin.

Some geological poetry was obviously whimsical, even satirical, but nevertheless had an important educational component, in its ability to help the reader understand (and even memorize) the sometimes overwhelming amount of information concerning the relationships between the various stratigraphical layers, rocks, and minerals. London-born John Scafe (1776–1843) began writing poetry while a student at Oxford, although he never graduated, and continued writing while serving nine years in the army. He began publishing his works in his forties, but did not see commercial success until the anonymously published *King Coal's Levee (1818),* a fanciful work inspired by *Outlines of the Geology of England and Wales* by William Phillips and William Conybeare. The work was embraced by the Geological Society of London in its efforts to popularize their science, Conybeare and William Buckland going so far as to aid Scafe in expanding the second and later editions through the inclusion of explanatory notes (O'Connor 2013). *The Monthly Review* termed it a "spirited little morsel, which is neither altogether a *jeu d'espirit* nor a lesson on the mineralogy of England, and which yet partakes of both…" (1821: 271).

An example of the poetic description of geological strata and processes is found in the verse introducing Granite, an igneous (fire-formed) rock that formed the primary (deepest) level of bedrock in many areas:

Duke Granite first; a hoary-headed sire,
Yet blest with symptoms of primaeval fire,
That beam'd across the traces of decay,
As vivid tints illume departing day. (Scafe 1820b: 4)

The accompanying geological note explains "Granite is the undermost (that we are acquainted with,) and therefore supposed to be the oldest of the *primitive* rocks" (Scafe 1820b: 46). The success of *King Coal's Levee* led to a likewise anonymously published sequel in 1820, *Court News; or the Peers of King Coal, and the Errants, or a Survey of British Strata.* The preface explains that "The Errants is wholly formed on the *Table of Order of Superposition of Strata*, by the Rev. W. Buckland, Professor of Mineralogy at Oxford, &c. placed at the conclusion of Phillip's *Geology of England and Wales*, from which latter work the Author has extended and endeavoured to elucidate the subject" (Scafe 1820a: np; emphasis added). In order to make it easier for the reader, the individual strata being referenced in *The Errants* are clearly labeled in the margin of each page.

A similar collection of works (also published anonymously) was *A Geological Primer in Verse*, attributed to Robert Bakewell by a reviewer in *The Monthly Review* (1821). The primer itself (only one of several individual works collected in this volume) was "intended as an Introduction to 'King Coal's Levee,' and is modeled on that erudite alphabetical effusion which begins with 'A was an Archer, and shot at a Frog,' &c." (*The Monthly Review* 1821: 275). It begins

> A was an Agate as round as a Ball.
> B was Basalt in the cave of Fingal.
> C was King Coal, of Oxford the pride.
> D Doubtful Diabase, close by his side. (Bakewell 1820: 21)

Another work in the same volume, "A Poetic Geognosy," hearkens back to *King Coal's Levee*, and describes the formation of the earth and the various geological strata by giving the rock types personalities corresponding to their properties and locations. It begins

> When Nature was young, and the Earth in her prime,
> All the Rocks were invited with Neptune to dine.....
> First Granite sat down, and then beckon'd his queen,
> But Gneiss stepp'd in rudely, and elbow'd between,
> Pushing Mica-slate further.... (Bakewell 1820: 7–8)

The volume also contains "Geological Cookery," humorous recipes for making rocks out of minerals. For example, "To make Granite./Of Feldspar and Quartz a large quantity take,/Then pepper with Mica, and mix up and bake" (Bakewell 1820: 37). A final component of the book is the essay "A Critical Dissertation on 'King Coal's Levee'." Here Bakewell urges the reader to first go through the geological notes found at the back of the work in order to understand the allusions present in the poetry. He specifically notes that in the case of the original publication of *King Coal's Levee*, although it was offered to its readers "without any meaning whatever… there are mischievous people ever ready to discover a meaning for it, and that frequently a bad one" (Bakewell 1820: 60). This refers to criticism of Scafe by those who interpreted his satirical comments about the stratification of the rocks and their hierarchies as a commentary against the Duke of Northumberland and other members of the royal class. After analyzing and criticizing the political interpretations of the poem for several pages, Bakewell dismisses them by "indulging a hearty laugh" (1820: 67).

While, according to Bakewell, Scafe did not have politics in mind, travel writer and poet Lady Emmeline Charlotte Elizabeth Manners Stuart-Wortley (1806–1855) most certainly did when writing her geological metaphor-laden poem "Reform-Liberty-March of the Intellect" (1837). Stuart-Wortley began publishing poetry in 1833, two years after marrying Hon. Charles James Stuart-Wortley (1802–1844), and continued to write while raising their three children. Overall, the quantity of her work is generally considered to be greater than its quality (Lee 2004). Much of her poetry drew upon her extensive travels, first throughout Europe, and after the death of her husband and youngest son in 1844, across the Atlantic into North and South America with her daughter Victoria. She also published prose volumes on her travels, including *Travels in the United States* (1851), *Sketches of Travel in America* (1853), and *A Visit to Portugal and Madeira (1854).* She died on a journey with her daughter to the Middle East in 1855. Five months after receiving a broken leg from a mule kick near Jerusalem, the mother/daughter pair and a maid unwisely set off to Aleppo without assistance. The maid died of sunstroke, Emmeline by dysentery, and Victoria barely survived (Robinson 1990).

According to John Robert Higgins (2011), "Reform-Liberty-March of the Intellect" uses geological processes as metaphors for political and social revolution, especially the French Revolution. Just as commoners can be unexpectedly lifted to the heights of society and power, lowlands can be unexpectedly thrust upwards to form mountains, leading to

> strange things said of their origin,
> Hints given of their low station in the past,
> Doubts cast on their *creation* and their *caste,*
> Whether they were of the Ancient Earth com-*peers*, [sic]
> Or of a later date—of modern years. (Stuart-Wortley 1837: 193–4; emphasis original)

Similarly, just as those born to royalty can find themselves reduced to the lowest classes through a turn of fortune, mountains are lowered through the great equalizing processes of erosion, weathering, faulting and landslides:

> In short, we've heard a tale so dark and strange,
> And so replete with unexampled change,
> As to level these proud Mountains brings,
> Straight ways, with base, flat, plain, low under things—
> Enough to strip them of their claims, and thrust
> These high Pretenders back to native dust. (Stuart-Wortley 1837: 194)

Less serious than Stuart-Wortley and invoking some of the playfulness of Scafe is Elizabeth Cobbold's 1823 poem "Poetical Letter to a friend." Elizabeth Knipe Clarke Cobbold (1765–1824) published works under all three of her last names, beginning with *Six Narrative Poems* at the age of eighteen. Her first husband, customs officer William Clarke (c.1731–1791), was about twenty five years her elder, and died three months into their marriage. She remarried soon after, to brewer and banker John Cobbold (1746–1835), but not before publishing a lengthy novel entitled *The Sword* under her first husband's last name. Her writing was thereafter interrupted by her familial duties, as stepmother to John's seventeen children from his first marriage and mother to the sons (two of whom died in childhood) and a daughter who died in infancy that they produced together (Jermyn 1825). She became renowned in local social circles for the intricate scissor-cut paper valentines with accompanying poems that she handmade for guests at her annual valentine party; many of these works of art survive to this day. The year after her death a collection of poems was published, many released for the first time, along with a biographical essay.

Cobbold was deeply interested in many aspects of natural history, including botany, entomology, geology, mineralogy, and conchology, and published her observations of the natural world. For example, her paper "On the Fasciola Hepatica" concerning a liver fluke was read to the Linnaean Society on April 17, 1810 (*The Philosophical Magazine* 1810). She and her children collected fossil shells on their estate, many of which she shared with paleontologist James Sowerby (1757–1822). In recognition of her contributions to his work, Sowerby named a variety of saltwater clam after her, *Nucula cobboldiae*,

> being desirous of commemorating Mrs. Cobbold, whose copious collection obtained with great industry, in company with several of the junior branches of her family, whom she delighted to inspire with a love for the works of nature, from the Crag pits of her own estate, evinces a degree of taste and zeal seldom met with. (Sowerby 1818: 178)

Cobbold's love for fossil taxonomy is reflected in the terminology-heavy poem "A Poetical Letter to a Friend." After romantically describing a river flowing from a cave, she muses

> Here science thoughtful, seeks to trace
> The changes mark'd on nature's face,
> And finds, in strange confusion hurl'd,
> The wonders of a former world!
> Dense, conic, belemnitic stones,
> Sharks fossil teeth, and horses bones,
> And most abundant store of shells
> Invite research at Holy-wells. (Cobbold 1825: 83)

The poem then shifts gears in a significant way, from a romantic description to a laundry list of fossil shell types, including "Arca, pentunculus; and more/That I could name to many a score!" After admitting that "some apology/Is due for all this dull conchology" she leaves her "shells to grace the cabinet" (Cobbold 1825: 84).

Romantic poetry was a forte of British poet Felicia Hemans. The daughter of an Irish merchant, Felicia Dorothea Browne Hemans (1793–1835) was born in Liverpool, England, but spent much of her childhood in north Wales, where she was home-schooled in languages (Latin, French, Italian, Spanish, Portuguese, and German) as well as music and drawing. As her sister and biographer, Harriet Mary Browne, notes, it was in Wales that she "imbibed that intense love of Nature which ever afterwards 'haunted her like a passion'" (Browne 1839: 32). She discovered the works of Shakespeare at age six, and had composed a poem about her love for Shakespeare by age 11 (Browne 1839). By the time she was fifteen she had published several volumes of poetry, and had been brought to the attention of Percy Shelley, who corresponded with the young girl—under a pseudonym—until her mother became involved. In 1812 she married Captain Alfred Hemans (b. 1781), a wounded veteran. While they had five sons together within five years, the couple separated in 1818, and she continued her successful career as a poet of note (including poems written for commissions), publishing over 400 poems in magazines and anthologies as well as her own twenty volumes. She also produced hymns, songs, and lyrics.

Known as the British Nightingale, her song "Pilgrim Fathers" was said to have been particularly beloved by famed New England poet Henry David Thoreau, who quotes from her work "The Voice of Music" in his literary notebook (Moseley

1978). She passed away in 1835 after a bout of scarlet fever. She is one of a number of authors whose death is lamented in William Wordsworth's 1835 poem "Extempore Effusion upon the Death of James Hogg," and is remembered today through the annual Felicia Hemans poetry prize at the University of Liverpool (Moseley 1978). Curiously, Lydia Sigourney had also been called the American Hemans especially after one of her poems was mistakenly included in an American edition of Hemans' work (Haight 1930). Jane Kilby Welsh's *Familiar Lessons in Mineralogy and Geology* referenced her, when Emily offers that "Mrs. Hemans' requisites for a good poet, she thought, were all that were necessary for possessing a taste for scientific pursuit—'an eye to see, a heart to feel, nature;' to which, however, ought to be added the love and reverence of the Great First Cause of all…." (1832: 147).

Her sister explained of Hemans that "Some of the happiest days the young poetess ever passed were during occasional visits to some friends at Conway, where the charms of the scenery, combining all that is most beautiful in wood, water, and ruin, are sufficient to inspire the most prosaic temperament with a certain degree of enthusiasm…" (Browne 1839: 43). On one of these trips (circa 1814–5) she composed two related poems about a friend who was devoted to geology. Her sister offers "they were never intended for publication, but were merely a *jeu d'esprit* of the moment, in good-humored raillery of the indefatigable zeal and perseverance of one of the party in his geological researches" (Browne 1839: 46). The first is "Epitaph on Mr. W----, a Celebrated Mineralogist," in which Hemans compares the effects on the environment of his enthusiastic digging to that of natural geological processes. For example,

> such traces of his power, He's left from Penmaenbach to Penmaenmawr,
> Such caves, and chasms, and fissures in the rocks,
> His works resemble those of earthquake shocks;
> And future ages very much may wonder
> What mighty giant rent the hills asunder,
> Or whether Lucifer himself had ne'er
> Gone with his crew to play at foot-ball there….
> to secure soft slumber to his bones,
> We paved his grave with all his favourite stones.
> His much-loved hammer's resting by his side;
> Each hand contains a shell-fish petrified:…. (Browne 1839: 46–7)

The second poem is "Epitaph on the Hammer of the Aforesaid Mineralogist," the faithful instrument said to lie in eternal repose beside its owner:

> Here in the dust, its strange adventures o'er,
> A hammer rests, that ne'er knew rest before.
> Released from toil, it slumbers by the side
> Of one who oft its temper sorely tried;
> No day e'er pass'd, but in some desperate strife
> He risk'd the faithful hammer's limbs and life;… (Browne 1839: 48)

The geologist is never identified.

Charlotte Turner Smith (1749–1806) was a poet and novelist. Her mother passed away when Charlotte was only 3 or 4 years old, so much of her upbringing fell to a maternal aunt, Lucy Towers. Smith excelled at acting at a girl's school in Kensington, where she also learned music, dancing, and drawing, and began writing poetry. While her father encouraged her to write poetry (as he had), her aunt was of the opinion that "learning disqualified women for their own peculiar duties" and actively discouraged her young ward's appetite for reading (*Monthly Magazine* 1807: 244–5). She was later schooled at home until her father accepted a marriage proposal from Benjamin Smith (six years her elder) when Charlotte was fifteen. Benjamin was the son of a director of the East India Company but disappointed his father, Richard, by demonstrating a lack of business acumen. Fortunately, his daughter-in-law Charlotte proved an asset, helping with the business' correspondence despite being quite busy birthing a dozen children (several of whom died in childhood). Their relationship became stronger when her maternal aunt became Richard's third wife. When Richard was libeled, Charlotte wrote a vindication; afterwards Benjamin was sent to the country to become a gentleman farmer, but financial troubles followed. Upon Richard's death in 1776 his will was found to be open to interpretation, despite the fact that he had tried to protect his grandchildren's inheritance from his son's demonstrated financial ineptitude. The result was that the estate was not settled in full until 1813, and Charlotte took up writing to support the family, including the funding of a settlement to commute her husband's seven month sentence in debtor's prison (Zimmerman 2004).

Originally publishing *Elegiac Sonnets, and other Essays* (1784) under the pseudonym Charlotte Smith of Bignor Park, Sussex, she later published nearly a dozen novels under her own name after the success of her original work (which grew to two volumes by the ninth edition, published in 1800). Her other works include translations, a two-volume history of England, and two posthumous works, *A Natural History of Birds* (1807) and *Beachy Head, Fables, and other Poems* (1807). She also wrote a number of books for children, "eminently calculated to form the taste, instruct the mind, and correct the heart" (*Monthly Magazine* 1807: 248). An example is *Rural Walks: In Dialogues*, (1795), a series of dialogues between the widow Mrs. Woodfield, her two teenage daughters, and, in an autobiographical turn that was common in her writing (and reminiscent of a similar turn in Jane Kilby Welsh's works [Sect. 5.2]), her orphaned niece as well (among other characters). Smith's stated goal was to address what she called "*les petites morales*," including the "flippancy of remark, so frequently disgusting in girls of twelve or thirteen" and "give them a taste for the pure pleasures of retirement, and the sublime beauties of Nature…." (Smith 1795: iii–iv). As descriptions of plants and the habits of animals are central to many of the tales, this work could be considered under the aegis of the familiar format of science writing. Her poetry was very popular in its day, and her preference for sonnets is said to have influenced John Keats and William Wordsworth. It is also suggested that she and Gothic novelist Ann Radcliffe (1764–1823) mutually influenced each others' writings (Stanton 2003).

Smith survived her husband (from whom she was separated and who spent the end of his life in a debtor's prison once again) by less than a year.

"Beachy Head" (1807), published posthumously, appears to be a fragment of a longer work, but some critics have argued that this form is intentional, and that the poem was actually complete (Higgins 2011). At the heart of the poem are two parallel mysteries, the geological history of the cliff, and the life history of a hermit who lives in a cave within the cliff. The poem begins with the speaker surveying the landscape from the top of the cliff:

On thy stupendous summit, rock sublime!
That o'er the channel rear'd, half way at sea
The mariner at early morning hails,
I would recline; while Fancy should go forth,
And represent the strange and awful hour
Of vast concussion; when the Omnipotent
Stretch'd forth his arm, and rent the solid hills,
Bidding the impetuous main flood rush between
The rifted shores, and from the continent
Eternally divided this green isle. (Smith 1807: 1–2)

The reference here is to one of two hypotheses of the time concerning the geological development of England as an island. The first, as Smith (1807: 143–4) explains in a note, is the "idea that this Island was once joined to the continent of Europe, and torn from it by some convulsion of Nature. I confess I never could trace the resemblance between the two countries. Yet the cliffs about Dieppe, resemble the chalk cliffs on the Southern coast. But Normandy has no likeness whatever to the part of England opposite to it." An alternative hypothesis, in which the island was thrust upward from the ocean floor, is mentioned later in the poem:

Or did this range of chalky mountains, once
Form a vast bason [sic], where the Ocean waves
Swell'd fathomless? What time these fossil shells,
Buoy'd on their native element, were thrown
Among the imbedding calx: when the huge hill
Its giant bulk heaved, and in strange ferment
Grew up a guardian barrier, 'twixt the sea
And the green level of the sylvan weald [sic] (Smith 1807: 26–7)

Smith also references her own personal experiences fossil hunting along the shore:

And still, observing objects more minute,
Wondering remark the strange and foreign forms
Of sea-shells; with the pale calcareous soil
Mingled, and seeming of resembling substance…. (Smith 1807: 26)

In a related note she explains further that "Among the crumbling chalk I have often found shells, some quite in a fossil state and hardly distinguishable from chalk. Others appeared more recent…. It is now many years since I made these observations. The appearance of sea-shells so far from the sea excited my surprise, though I then knew nothing of natural history" (1807: 158–9). The paralleling of this incident with the opening tableau of Maria Hack's *Geological Sketches* (written 25 years later) is interesting, suggesting a possible influence.

In the end, neither the life of the hermit nor the cliff can be known with certainty, and Noah Heringman (2004) argues that this is meant as a lesson to the geological community.[13] As Smith herself warns,

Ah! very vain is Science' [sic] proudest boast,
And but a little light its flame yet lends
To its most ardent votaries; since from whence
These fossil forms are seen, is but conjecture…. (Smith 1807: 27)

In her notes she offers further "I have never read any of the late theories of the earth, nor was I ever satisfied with the attempts to explain many of the phenomena which call forth conjecture in those books I happened to have had access to on the subject" (Smith 1807: 159). It would be interesting to know which books she was describing, but given the date of her poem it would not have been the popularizations of geology written by the women featured in this present volume.

Hannah Flagg Gould (1789–1865) was born in Lancaster, Vermont but spent the majority of her life in Newburyport, Massachusetts. After her mother's death she became the constant companion to her father Benjamin (a veteran of the Battle of Bunker Hill and other Revolutionary War battles). Her father's experiences influenced her poetry, which first appeared in popular periodicals before publishing collections of her own in 1832, 1835 and 1841. A prose collection, *Gathered Leaves*, was published in 1846. The volume *New Poems* appeared in 1850. Her writings had a decidedly Christian flavor that was considered appropriate for family reading (Cleveland 1859). Interestingly her elder brother Benjamin Apthorp Gould's son of the same name became an astronomer, receiving a Ph.D. from Göttingen after studying under Carl Gauss (Wilson and Fiske 1900). This was the same gentlemen who was influential in introducing the term *scientist* into the American vernacular (Sect. 1.1).

Two geology-influenced poems appear in *New Poems*. The first, "The Opal," describes both the distinctive way that the stone interplays with light and its cultural history, and then subsumes both into a metaphor for the relationship between a Christian and God. The poem is preceded by a short introduction summarizing opal's color, history, and various superstitions before the evanescent stone is described in the poem proper as a

Gem with the burning heart,
That, like a living soul,
Pervading every part,
Gives beauty to the whole. (Gould 1850: 57)

After revisiting some of the superstitions in Rome and the Orient she offers

But to our Christian sight,
While traced all over thee
Thy Maker's name, as light,
Shines forth "the Deity,"
On *Him*, our Rock,- the "Living Stone,"-

[13] For more analysis of the geology of "Beachy Head," see Heringman (2009).

Through faith and hope, we rest alone.
For He who formed thee thus,
So curious, rich and fair,
Hath still assigned to us
His holier seal to bear,-
His *image*, that we may be shown,
As priceless jewels, marked his own. (Gould 1850: 58; emphasis original)

A second poem, "The Mastodon," was originally published in a magazine in 1847. In this work she paints the mysterious, slumbering behemoth as crudely awakened by modern humans in the age of the steam engine:

Thou ponderous truth, from thy long night's sleep
Through the unrecorded eras
Awaked, and come from their darkness deep
To this day of light chimeras!-
What wast thou, when thy mountain form
Stood forth in vital glory?
O, who can paint thee live and warm,
Or reveal thy life's strange story?...
Age on age, with their stone and mould,
In strata deep, then made thee
A shroud no power could e'er unfold,
Till a day of *steam* betrayed thee…. (Gould 1850: 80; emphasis original)

It is interesting to contrast this depiction of the beast to Godding's statement that "his history is veiled in obscurity. We only know that such a noble animal once walked the earth, and that its massive bones, after many centuries had passed away, have come forth at last from their silent resting place, to speak to us of the wondrous power and wisdom of the Creator" (1847: 74). In Godding's description we see that not only is geology often the inspiration for poetry, but that geological prose often contains something of the poetic within it, in its ability to inspire the reader to contemplate events of almost inconceivable antiquity.

References

Aikin J (1777) An essay on the application of natural history to poetry. Johnson, London
Bakewell R (1820) A geological primer in verse. Longman, Hurst, Rees et al., London
Browne HM (1839) Memoir of the life and writings of Mrs. Hemans. Lea and Blanchard, Philadelphia
Burrill K (1904) School-books old and new. Chamber's J 8:6–9
Church Review and Ecclesiastical Register (1848) First lessons in geology. By Miss D.W. Godding. 1(2): 288
Cleveland CD (1859) A compendium of American literature. Biddle and Biddle, Philadelphia
Cobbold E (1825) Poems of Mrs. Elizabeth Cobbold. Raw, Ipswich
Collections of the Minnesota Historical Society (1880) First annual circular and catalogue of the Sigourney Boarding School, St. Anthony, Minnesota. 1860–1. 3:52
Collin GL (1902) Lydia Huntley Sigourney. N Engl Mag 27(1):15–30

Connecticut Historical Society. Delia Godding correspondence, 1847–1855. Ms 101610, Folder 1 of 1
Darwin E (1798) The botanic garden. Swords and Swords, New York
Emery C (1941) Scientific theory in Erasmus Darwin's 'The botanic garden'. Isis 33(3):315–325
Fawcett LM (1933) Frontier education. Minn Hist Mag 14(2):142–149
Godding DW (1847) First lessons in geology. Parsons, Hartford
Godding R (1867) East Burke. Vermont Historical Gazetteer 1:210–311
Gould HF (1850) New poems. Reynolds, Boston
Greenlee RS, Greenlee RL (1908) Genealogy of the Greenlee families. Greenlee, Chicago
Greer (1857) Greer's Hartford City Directory for 1857–8. Greer, Hartford
Haight GS (1930) Mrs. Sigourney: the sweet singer of Hartford. Yale UP, New Haven
Hartford Daily Courant (1844) Classified ad. Dec 21:3
Hartford Daily Courant (1852) List of premiums: awarded by the Hartford County Agricultural Society. Oct 25:2
Hemenway AM (ed) (1867) Delia Godding. Ver Hist Gaz 1:312–313
Heringman N (2004) Romantic rocks, aesthetic geology. Cornell UP, Ithaca
Heringman N (2009) 'Very vain is science' proudest boast: the resistance to geological theory in early nineteenth-century England. In: Rosenberg GD (ed)The revolution in geology from the Renaissance to the Enlightenment. Geol Soc Am Mem 203:247–257
Higgins JR (2011) Fossil poetry, the birth of geology, and the romantic imagination, 1790–1860. Dissertation, University of South Carolina
Hitchcock E (1845) Elementary geology, 3rd edn. Newman, New York
Hitchcock E (1851) The religion of geology and its connected sciences. Collins, Glasgow
Jermyn L (1825) Memoir of Mrs. Elizabeth Cobbold. In: Jermyn L (Ed) Poems of Mrs. Elizabeth Cobbold. Raw, Ipswich, p 3–42
Lee E (2004) Wortley, Lady Emmeline Charlotte Elizabeth Stuart- (1806–1855). Rev. Gill Gregory. Oxford Dictionary of National Biography, Oxford UP, Oxford. http://www.oxforddnb.com/view/article/26728. Accessed 4 Apr 2017. doi:10.1093/ref:odnb/26728
Monthly Magazine (1807) Memoirs of eminent persons: Mrs. Charlotte Smith. 23: 244–248
Moseley C (1978) Henry David Thoreau and Felicia Dorothea Hemans. The Concord Saunterer 13(2):5–8
Niles J (1864) Diary. http://www.usgennet.org. Accessed 17 Mar 2017
O'Connor R (2007) Young-Earth creationists in early nineteenth-century Britain? towards a reassessment of 'scriptural geology'. Hist Sci 45:357–403
O'Connor R (2013) The earth on show. Chicago UP, Chicago
Ogilvie M, Harvey J (eds) (2000) The biographical dictionary of women in science, vol 1. Routledge, New York
Parley P (1857) Thousand and one stories of fact and fancy, wit and humor, rhyme, reason, and romance. Derby and Jackson, New York
Primer I (1964) Erasmus Darwin's Temple of nature: progress, evolution, and the Eleusinian mysteries. J Hist Ideas 25(1):58–76
Robinson J (1990) Wayward women: a guide to women travelers. Oxford UP, Oxford
Scafe J (1820a) Court news; or, The peers of King Coal: and The Errants; or A survey of British strata. Longman, Hurst, Rees et al., London
Scafe J (1820b) King Coal's levee or geological ettiquette, 4th edn. Longman, Hurst, Rees et al., London
Sigourney L (1862) The man of Uz and other poems. Williams, Wiley and Waterman, Hartford
Smith C (1795) Rural walks: in dialogues. Wogan et al, Dublin
Smith C (1807) Beachy Head with other poems. Johnson, London
Sowerby J (1818) The mineral conchology of Great Britain, vol 2. Sowerby, London
St. John's Parish (n.d.) http://www.sjparish.net/About_us/Church_history/. Accessed 2 Apr 2017
Stanton JP (ed) (2003) The collected letters of Charlotte Smith. Indiana UP, Bloomington
Stuart-Wortley E (1837) Impressions of Italy. Saunders and Otley, London

The Kosciusko Star (1894) The eminent dead—death of Judge Jason Niles last Saturday. July 13. 1:8
The Monthly Review (1821) King Coal's levee, or geological etiquette. 94: 265–277
The Philosophical Magazine (1810) Linnaean Society. 35:384–385
Welsh JK (1832) Familiar lessons in mineralogy and geology, vol 1. Clapp and Hull, Boston
Willson CC (1894) Edgar P. Inglee vs. Hebert T. Welles et al. Minnesota Reports v 53: 197–198
Wilson JG, Fiske J (1900) Gould, Benjamin. In: Appletons' Cyclopædia of American biography, vol 2. Appleton, New York, p 693
Zimmerman SM (2004) Smith, Charlotte (1749–1806). Oxford Dictionary of National Biography. Oxford UP, Oxford. http://www.oxforddnb.com/view/article/25790. Accessed 4 Apr 2017. doi:10.1093/ref:odnb/25790

Chapter 8
Almira Hart Lincoln Phelps (1793–1884): Textbooks and the Female Seminary

8.1 Biographical Overview

As illustrated by the life of Delvalle Lowry and others (Sect. 6.3), the influence and support of a familial mentor had the power to change lives, especially in a culture in which the scientific contributions of women were particularly underappreciated, and opportunities therefore limited. A special case is when the mentorship involves an older sister who shares a young woman's passion for science, writing, or more generally education. This camaraderie in the trenches, metaphorically speaking, made all the difference in the life of Almira Hart Lincoln Phelps.

Almira Hart (1793–1884) was born in Berlin, Connecticut, on July 15, 1793, the youngest of 17 children fathered by farmer Samuel Hart and the tenth born to his second wife, Lydia Hinsdale. Almira was very close to her sister, Emma (1787–1870), who was closest to her in age. Known for their liberal politics, the Harts encouraged their daughters to seek out educational experiences and kept their house well stocked with books (Bolzau 1936). Both girls attended the Berlin district schools before receiving additional secondary education elsewhere. In 1812 Almira attended an academy in Pittsfield, Massachusetts, run by her cousin, Miss Nancy Hinsdale, and where Emma had earlier taught. She returned to her hometown at the age of 14 and became a teacher herself, first at a Berlin school and then in neighboring New Britain. Henry Barnard, first principal of the New Britain Normal School, remembered that this was the first time that a "female teacher was placed in charge of the winter school of a large district," and described how she handled herself in a particularly awkward moment in her job interview:

> In her examination as a candidate, she covered her embarrassment caused by her failure to answer with minute accuracy a question as to the "exact distance of the largest fixed star from the planet Mars," by reading an original essay "On the Duties and Responsibilities of the Teacher," in which she at once exhibited her technical knowledge of reading, writing, and orthography, and her appreciation of the office for which she was on examination. (Barnard 1868: 612)

K. Larsen, *The Women Who Popularized Geology in the 19th Century*,
DOI 10.1007/978-3-319-64952-8_8

Emma Hart had meanwhile continued her own teaching career, including becoming the preceptress at a female portion of the academy in Middlebury, Vermont. She temporarily gave up her career in 1809 to become the second wife of Dr. John Willard, a widowed physician nearly 30 years her senior and father to four children from two prior marriages. Emma gave birth to her only biological child, John Hart Willard, in 1810, and Almira moved in with the family for the next 2 years. While in Middlebury Almira was able to listen to lectures at the Middlebury Academy (later Middlebury College), and afterwards returned to Nancy Hinsdale's Academy at Pittsfield, Massachusetts for more formal education and teaching experience. Emma meanwhile opened a school in her home after her husband had suffered significant financial losses. These were caused by a robbery at the Vermont State Bank, which he then served as Director (Grigg 1999). Over the next decade Emma began to establish her reputation as a tireless advocate and vocal proponent for women's education, first in Vermont and then in New York State.

Almira was following a parallel trajectory back in Connecticut, albeit at a slower pace. In 1814 she opened her own female boarding school in Berlin, and soon after became the principal at a school in Sandy Hook, New York. She temporarily gave up her career as an educator on October 4, 1817 when she married Simeon Lincoln, editor of the *Connecticut Mirror*, in Hartford. During the next 6 years she bore Simeon three children, two of whom, daughters Jane Porter and Emma Willard, survived. There were also upheavals in their lives. In late 1821 Simeon's father was lost at sea in a shipwreck, and he resigned from the newspaper to take over his father's farm in New Britain, but died of Yellow fever in 1823 (Bolzau 1936). In order to support her family, Almira taught briefly in New Britain before moving to Troy, New York in 1824 to become a teacher at the Troy Female Seminary, which Emma had opened in 1821. It was most probably here that Almira met Amos Eaton, who gave lectures at the female academy and tutored the female teachers in order to raise their own level of proficiency. After becoming Senior Professor at the new Rensselaer School (later the Rensselaer Polytechnic Institute) Eaton still kept close ties with the female seminary and some women attended his lectures at Rensselaer.

Almira was given the responsibility of teaching various science topics, including botany and chemistry, and she had to develop (with Eaton's aid) her own syllabi and lectures. Her lecture notes on Botany became her most famous and successful work, *Familiar Lectures on Botany* (1829). In order to help her own students, she also translated into English and revised (at Eaton's urging) Louis-Nicolas Vauquelin's *Dictionnaire de Chimie*, published in 1830 (Bolzau 1936). Almira also served as vice-principal of the school for a number of years, and temporarily as principal when Emma travelled to Europe in 1830–1. Emma returned to find that Vermont Judge John Phelps (1777–1849), whose daughter, Lucy, was a student at the seminary, had proposed to her sister. Almira and John Phelps were married on August 17, 1831, and Almira and her younger daughter Emma moved to Guilford, Vermont to make a home with the Judge and his two younger children, Lucy and Regina Ann. According to Almira's biographer, several of the Judge's four elder children and Almira's elder daughter, Jane, did not initially approve of the marriage (Bolzau 1936).

The Phelps family moved from Guilford to the larger town of Brattleboro, which featured a lyceum that hosted public lectures. For example, a January 16, 1835 lecture on "The principles and facts of modern Geology, confirmatory to the Mosaic account of the creation of the world" was reported in the *Vermont Phoenix* (Bolzau 1936). With her husband's support, Almira wrote, edited, and revised a number of works, mainly textbooks based on her time at the Troy Female Seminary—*Familiar Lectures on Natural Philosophy* (1837) and *Familiar Lectures on Chemistry* (1838)—as well as simpler versions of these texts (and *Familiar Lectures on Botany*) for younger students. She also revised and expanded a geology text for young children, *The Child's Geology* (1832). Like her sister, Almira had become deeply passionate about women's education, and published more generally on this issue. One of these works, *Lectures to Young Ladies, or The Female Student* (1833) was so popular that it was adopted by the Massachusetts Board of Education for *The School Library* in 1838 (as had the unauthorized revision of Jane Marcet's *Conversations for Children; on Land and Water* [Sect. 3.3]). Almira also attended to the education of her youngest daughter and stepdaughter, offered lectures to local teachers and young women, and found the time to give birth to two more children, Charles Edward and Almira Lincoln. The Phelps' were also active in their community, especially in their local church. All of Almira's writings reflect a deep faith and the conviction that the study of the natural world is a highly effective pathway to realizing the grand design of God's plan for the universe (a theme that is discussed further in Sects. 8.2, 8.3 and 8.4).

While Almira's fame grew, thanks to the sales of her various science books, John Phelps entered politics, serving as a State Senator but resigning the post after a year because he refused to join the popular Abolitionist party due to his personal convictions concerning the sanctity of states' rights (Bolzau 1936). The Phelps' accepted the opportunity to leave Vermont when Almira was offered the post of principal at the newly built West Chester Young Ladies' Seminary in Pennsylvania. The school failed after several years due to financial difficulties, and the Phelps' founded the Rahway Female Institute in New Jersey in 1839. Two years later they accepted an invitation to travel to Ellicott's Mills, Maryland, to start a new school, the Patapsco Female Institute.

John Phelps died on April 14, 1849, and Almira retired in 1856 after the death of her eldest daughter, Jane, in a railroad accident, moving to Baltimore (Bolzau 1936). Her last publications focused more on religion than science, including *The Christian Household,: a donation to the Baltimore Church Home* (1848) and *Hours with my Pupils, or The Educator* (1858). She also contributed to the *Church Quarterly Reviews* and in 1866 she wrote a paper on the relationship between science and religion in the works of Edward Hitchcock. This was read at a meeting of the American Association for the Advancement of Science. Almira had become only the second female member of this organization, after astronomer Maria Mitchell (Barnard 1868). She died on her 91st birthday, in Baltimore, Maryland, after an illustrious career that focused on science education.

8.2 Lectures to Young Ladies (1833)

Principal Emma Hart Willard had established a custom at the Troy Female Seminary to gather the students together on Saturday mornings "to give general instruction on subjects connected with literature, morals and religion" (Phelps 1833: 14); Almira continued this tradition while substituting for her sister in 1830–1, and later published her lectures in book form as *Lectures to Young Ladies* (1833). The general purpose of the lectures, as described by Almira in her introduction, was

> to awaken in the minds of the pupils, habits of thought with respect to the nature and design of education, and the practical application which ought to be made of its various branches. The young ladies to whom they were addressed were pursuing studies of different kinds; and it was important that they know the '*why* and the *wherefore*,' with the broad and general principles of literature and science. (1833: 6)

In a more modern vernacular, Phelps was attempting to explain to her students why they were receiving a liberal arts and sciences education. To the girls themselves, she explained that she would connect the various topics of their general education "with their practical application, female manners and accomplishments, the peculiar duties of woman in her domestic relations, towards society, and towards her Maker" (1833: 14).

Phelps' initial admonition to her charges sets the tone for the remainder of the book (written in the style of a solemn, didactic sermon spoken in first person as if the reader were one of her students sitting there on a Saturday):

> I see in the gay, unreflecting girl, the future wife, the mother, and the candidate for immortality, having power not only over her own destiny, but capable of wielding vast influence over other immortal beings. Impressed as these solemn truths are upon my mind, I may at times, seem to expect from you too much seriousness and reflection, at an age when gaiety and thoughtlessness are so natural. (1833: 15)

Phelps makes it clear to these young women that she sees their ultimate place in society as not merely fulfilling the stereotypical female domestic roles, but as intelligent, persuasive, and above all else, devoutly religious citizens. These were the myriad roles that their well-rounded seminary education was preparing them for.

Science education might have seemed a frivolous luxury when taken at face value. After all, there was a very practical reason why the common term was *gentlemen of science*; opportunities for women in science at this time were, as we have seen, extremely conscripted and limited. But it was proper for women to become teachers, collect minerals and fossils for their family's natural history cabinet, and be knowledgeable about the plants in the family garden (including medicinal herbs). Some might go so far as to become a scientific handmaid or amanuensis to a male relative. There was also one further proper role for ladies that would necessitate a passing knowledge of science, one that Phelps in particular acknowledges to her young charges:

> Now all that I plead for, is, that you may have enough of science to lead you to wish for more; enough to render you interesting companions to men of science. They will not generally expect more of you, than that you should be attentive listeners, or be able to suggest subjects for their explanation (1833: 214).

One hopes that Phelps paused in her lecture for emphasis before delivering the next line:

> Yet should you chance to become sufficiently acquainted with any branch of science to enable you to impart information, I know of no law, either of morality or propriety, which would be violated by your modestly imparting that knowledge to others; neither do I think any man of real science would be displeased to find a lady capable of supporting conversation on scientific subjects (Phelps 1833).

The relationships of the Murchisons, Marcets, and other famous nineteenth century power couples of science clearly demonstrate the truth in these words.

The above passage appears in Lecture18, which focuses on mineralogy and geology. Its scientific content includes a summary of the various types of minerals and common rocks, as well as their properties and practical uses, and an overview of the major strata. Attention is not given to the *processes* involved in geology. Phelps is very clear to explain to the young ladies that the study of mineralogy should be more of an afterthought, and not take away from time that is necessary for other studies. However, she offers that "very little attention to the subject, especially after you have become acquainted with chemistry, will enable you to comprehend the general features of the science, and will render cabinets of minerals something more to you than collections of glittering stones" (Phelps 1833: 214). She explains that in assembling their mineral cabinets, it is necessary to use some system of classification, and recommends that of American mineralogist Parker Cleaveland, whose influential 1816 textbook *An Elementary Treatise on Mineralogy and Geology* was in its second edition (1822). She urges them to be diligent in collecting their own samples, as "Every walk and every journey, may present you with something for your cabinet, and a new interest will thus be given to the face of nature" (Phelps 1833: 217), a common refrain in these works.

Phelps reiterates that "To females, geology is chiefly important, by its effect in enlarging their sphere of thought, rendering them more interesting as companions to men of science, and better capable of instructing the young" (1833: 218). But as in the case of many of the women science writers of that century, Phelps believes that "Especially does geology afford important aid to religion by confirming the truth of revelation... that as the structure of the earth is investigated, and the secrets of its interior brought to light, the strictest coincidence is observed between them, and the facts recorded in Scripture" (Phelps 1833: 218). She quotes in support of her views Benjamin Silliman of Yale's opinion that "the period is not far distant, when geology will be admitted into the train of her elder sister, astronomy, and that both will be eventually hailed as the friends and allies of revealed religion" (Phelps 1833: 218–9). However Phelps is among that subset of writers who does not stop at an alignment between geology and Scripture, but embraces the more conservative view that geology provides proof of the literal veracity of Scripture. As she states, "The physical history of the Deluge is everywhere inscribed upon the surface of the earth; upon its chasms and cliffs, its valleys and mountains. For a knowledge of the moral cause of these convulsions, we must look to the Scriptures" (Phelps 1833: 219). After quoting from the Bible concerning God's punishment of the wicked in the

Deluge, she proclaims "This one grand proof of the Scriptures, offered by geological science, is enough to entitle it to the attention of the Christian, for it furnishes sensible demonstration, broad and stable as the earth, of the truth of that book, which traces man from his creation and first planting upon this globe, and carries him into eternity" (Phelps 1833: 219).

While she takes a rather religiously literal view of the geological record, she aligns with the majority of writers in taking a metaphorical view of the six days of creation, in line with Bakewell's (and others') interpretation of them as six lengthy periods of time. But this appears to be the limit of what she is willing to accept as accommodation in order to align science and Scripture. She then returns to her original claim of geology's ability to provide proof for the historicity of Genesis, and in particular cites the diluvium as clear evidence of the Mosaic flood, and uses the order of fossil life found in the strata to confirm the order of events of creation in Genesis leading up to the Deluge (as had Jane Kilby Welsh [Sect. 5.2]). As she proclaims to her students, "Now, compare these facts as revealed by geology, with the events recorded in scripture, between the 1st and 10th chapters in Genesis, and you will find an entire corroboration of what is there recorded" (Phelps 1833: 234).

While it can be argued that *Lectures to Young Ladies* is not an excellent example of the familiar format, it does marginally fit within the genre. Phelps, the female authority figure, is having an intimate (albeit one-way) Saturday conversation with her female students in the domestic setting of their female seminary. She is cognizant of her audience's needs and abilities (as she openly defines these for the girls) and accordingly adapts her content to her audience, telling them what, in her mind, they need to know about the science and no more. The integration of morality and religion is certainly front and center in her writing. The only debate is whether or not there is at least lip-service paid to active learning and everyday examples. It is argued that her urging of the girls to collect samples while on daily walks for the purpose of arranging a mineral cabinet minimally meets this criterion. One might also argue that, to Phelps, there plainly is no greater or more obvious everyday example than Scripture; her argument that geological evidence reflects the events in Genesis is the integration of everyday examples into a discussion of science.

Regardless of whether or not one agrees with this interpretation, these religious views were undeniably central to Phelps' view of science, and had a great impact on both her original textbooks and her works for younger children. As will be demonstrated, they also allow scholars to reveal the ink from her pen in the work of another author.

8.3 The Child's Geology (1832)

Three of Almira Phelps' attributed writings are shrouded in some level of mystery; coincidentally, all relate to geology as well. The first is a work that was apparently never published; it will be considered alongside its cousins in the next section. The second is a work by an unknown author that she revised and enlarged, *The Child's*

Geology (1832), while the third is a set of potentially unattributed sections within the famous *Universal Geography* by William C. Woodbridge and Emma Willard (1824). This last work has a particularly interesting genesis. While teaching at her private school in Middlebury, Emma Willard had developed a new pedagogical strategy for teaching geography, one that abandoned the idea that geography should be memorized as a set of facts about each country or region and instead focused on more general concepts (an understanding of which could then be applied to the comparison of individual geographical examples). This became the geography curriculum taught in the Troy Female Seminary. Independently, William C. Woodbridge, a teacher at Thomas Gaullaudet's American Asylum for the Education of Deaf and Dumb Persons in Hartford, Connecticut, had developed his own similar system of teaching geography (Calhoun 1984). When Woodbridge learned of Willard's system in 1820, it was decided that they would issue a joint publication under the title *Universal Geography*, with Woodbridge's name affixed to the longer and more prestigious Modern Geography section and Willard listed as the author of the Ancient Geography portion.[1]

According to Henry Barnard, Almira Phelps' "first attempt at writing for the press was the preparation, at the request of the authors, of the chapters on Geology, and Roads and Canals, for Woodbridge's and Willard's Geography" (1868: 616). If this was indeed the case, then this provides additional samples of Phelps' writings on geology. The geological summary in *Universal Geography* is fairly standard fare for the times, an overview of the main strata and the common rocks found in each. However, a comparison of the style and terminology used in the geology summary in *Universal Geography* (1824 edition) with that of Phelps' mineralogy and geology lecture in *Lectures to Young Ladies* shows little similarity. There are similarities between a discussion of primitive rocks in *Universal Geography* and another of Phelps' troublesome works, *The Child's Geology*, as shown in comparison below:

> The *materials* of primary rocks... form a soil which is not well adapted to the production of vegetables.... The *rivers* of primary countries roll over rocky beds, full of obstructions, and rarely admit of any continued navigation (*Universal Geography* [Woodbridge and Willard 1824: 20; emphasis original]).
>
> Countries in which the primitive rocks prevail are not easily cultivated; the rivers are broken by falls and navigation obstructed (*The Child's Geology* [Phelps 1832: 34]).

Two nearly identical diagrams are found in both works as well, one a diagram of the strata and the other a diagram of the overall structure of the earth (Fig. 8.1). However, as was seen in the case of Godding and Hitchcock, it was common at the time to borrow illustrations from other works, sometimes (as in this case) without attribution.

While Barnard's claim that Phelps was a contributor to *Universal Geography* cannot be definitively corroborated by a comparison of that work to either Phelps' *Lectures to Young Ladies* or *The Child's Geology*, a stylistic comparison of this last work to others bears interesting fruit. This slender volume of around 130 pages was published in 1832. The title page bears the intriguing attribution "By the author of

[1]The history of this very uncomfortable collaboration cannot be summarized in a few brief sentences. For more information, see Calhoun (1984).

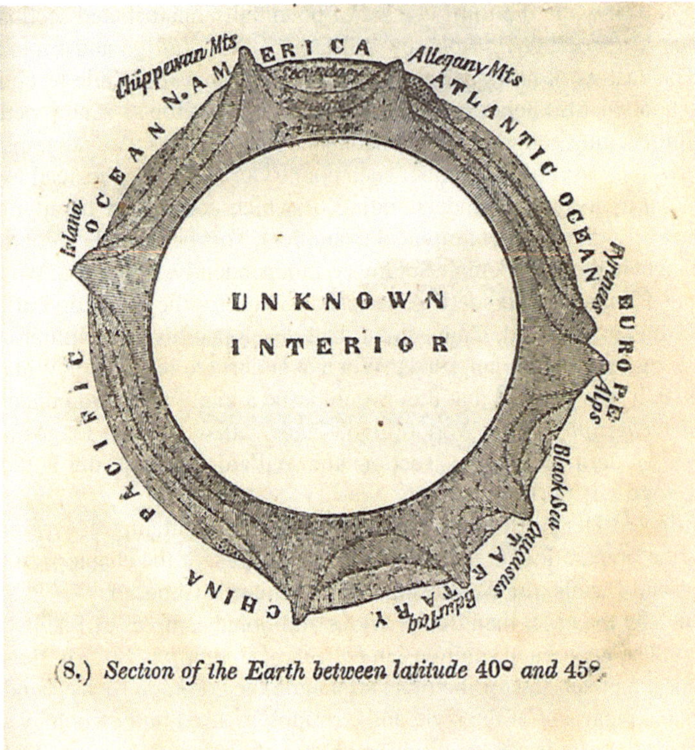

Fig. 8.1 Diagram of the earth's structure. Reproduced from William C. Woodbridge and Emma Willard (1829) *Universal Geography,* 3rd edn

The Child's Botany, Revised and Enlarged by Mrs. Almira H. L. Phelps." An advertisement for the work explains that it

> is offered as a guide to the young pupil in commencing the study of Geology. It is designed for Common Schools and the younger members of Higher Schools. It is also thought that a book of this kind, leading the child to view the "handy works of God," may serve a more valuable purpose for Sunday School Libraries than many of the fictions which are now to be found on their shelves. (Phelps 1832: 3)

The following page is a "Letter to the Children of the United States," written by Phelps, in which she gently breaks some difficult news to her young readers:

> My Dear Young Friends, A part of this book was written for you by the same person, who, a few years since, published a neat little volume called the Child's Botany. He was one who took a deep interest in the welfare of the young, as may be seen in his attempts to lead them to think of the Great Maker of all things.... Your kind and pious friend is now where you and I shall soon be – in eternity! While yet in the morning of his days, he was called from his earthly labours to enter, as we have reason to believe, "into that rest which remaineth for the people of God." If we study the *works* of our Heavenly Father with an [sic] humble sense of his power and wisdom, our minds will become more and more fitted to behold the fuller radiance of his glory which will burst upon those happy souls who shall be admitted

> into his presence. May you and I thus meekly and attentively study the character of God as manifested in the *works of His hands*, and confirmed in his *most Holy Word*. Your affectionate friend, Almira H.L. Phelps. (Phelps 1832:4; emphasis original)

The familiar tone and signature harken back to not only Phelps' *Lectures to Young Ladies*, but the works of Margaret Bryan. The religiosity reflected here is also characteristic of Phelps' other works.

The only questions are twofold—who is the man being eulogized by Phelps, and exactly what is the extent of the changes that Phelps brought to this volume in completing it for publication? *The Child's Botany* (1829) is commonly attributed to Samuel Griswold Goodrich (1793–1860), publisher of children's books and prolific author of many volumes himself under the pseudonym Peter Parley.[2] As Goodrich himself noted in his autobiography *Recollections of a Lifetime, or Men and Things I Have Seen* (1857), numerous plagiarists had benefited from his fame by publishing books under his pseudonym. In addition, because he published series of books by other authors, some of these volumes had become attached to his name. *The Child's Botany* is only referred to in a footnote as among the minor works that he had published but not written, and unfortunately he does not name the true author. For the sake of citations, therefore, the name of the text itself will be used, in order to not promulgate further the erroneous attribution to Goodrich.

While the first question must remain unanswered at the present time, a comparison of the text of *The Child's Botany*[3] to *The Child's Geology* (and both works to Phelps' other writings) gives a rather clear indication. As previously noted, there are some similarities between at least one section of *Universal Geography* and *The Child's Geology*, and at least two diagrams, so at the very minimum a one-way borrowing by Phelps of material from the earlier collaboration between her sister and Woodbridge is certain.

The Child's Botany begins with a preface explaining that.

> The writer of the following pages was induced to make an attempt at simplifying the first principles of Botany, for the use of children, from the belief that their minds are capable of understanding, and being interested in, many of those subjects which have generally been treated of in a manner solely adapted to the minds of mature age (*The Child's Botany*. 1831: iii)

The author then explains that while animal books are usually the first given to children, the topic of his book should also prove "useful and interesting to them" even if some parental encouragement is necessary to initially convince the child to read it (*The Child's Botany* 1831). The beginning of the preface to *The Child's Geology* starts in a similar enough tone:

> The object of the writer has been to present the leading facts of Geological science, in a form adapted to the understanding of children. How far he has been successful it remains for others to judge. But it is believed, that a *simple* history of the *internal structure* of the earth, and of the changes that have taken place on its surface, may be highly interesting and instructive *even* to children. (Phelps 1832: 5; emphasis original)

[2] This attribution appears in citations within the National Union Catalog, for example.

[3] The earliest edition that was available for comparison was the 4th (1831).

The second paragraph to this introduction also sounds suspiciously like Phelps's writing, especially in its message:

> The study of Geology has also a moral influence. It confirms the Mosaic account of the creation, notwithstanding the doubts of sceptics [sic]. It proves the account of a general deluge, the effects of which are everywhere visible; and, in fine [sic], irrefutably shows the harmony of the works of God with the records of revelation. (Phelps 1832)

The styles of the two works are also different in overall tone. *The Child's Botany* is written in very simple language and with a very informal first person perspective, and at about 100 pages is shorter than Phelps' work. The engaging style of the text can be seen from its first page:

> Little children! I intend to tell you a great deal about them [plants], in this book which I am writing, on purpose to instruct and amuse you. Perhaps you will think it is not so pleasant to read about trees and flowers, as to read stories or to go to play. But if you wish to live useful lives, and be respected and happy, you must read and study for instruction, as well as amusement.
>
> Well, how many different kinds of plants do you think there are in the world? More than thirty thousand have been discovered! (*The Child's Botany* 1831: 5-6)

The book then describes the parts of plants, followed by an overview of the Linnaean classification system with common examples. The widespread uses for each plant and other important properties are also included. The language continues to be simple and personal; for example in a description of roses the author notes "I have seen three of the above kinds of roses growing on one bush" (86). While there is a moral undercurrent (the betterment of oneself through learning), the work has very few religious references. For example, in speaking of grass, trees, and flowers, the author proclaims simply "God has given us these things, because they are very pleasant to our eyes" (*The Child's Botany* 1831: 12). There is also no overall conclusion, just an alphabetical list of the common plants of New England. There are questions included as footnotes to each page that can be used by parents to direct their child's attention, as suggested in the preface.

The Child's Geology also features this type of questions, but this is one of the few similarities. The book includes a discussion of the general form of the earth and its creation, fossils, the interior of the earth, strata, volcanic rocks, earthquakes and volcanoes, and then a summary of the geology of the United States and other countries. There is also a summary of the history of geology from the seventeenth century, including the Neptunian/Plutonian debate. A table summarizing the minerals and rocks referenced in the text is also included as an appendix. An explanatory note offers that the goal is for the reader to

> become familiar with the classification, and be enabled, if so disposed, to arrange the specimens in a cabinet. Specimens of all these minerals and rocks may be easily obtained; the whole number does not amount to 40. Every teacher of a common school should be provided with a geological cabinet. The scholars should be taught to examine the rocks and stones which they meet with in their daily walks. There should be a depository for minerals in every school-house, and the scholars encouraged to contribute – even the poorest offering should be encouragingly received by the teacher. (Phelps 1832: 126–7)

This instruction to collect samples from one's daily walk is suspiciously similar to that given by Phelps to the young ladies at the Troy Female Seminary in her Saturday lecture.

But the inclusion of one further topic undeniably demonstrates the influence of Phelps' pen: Genesis and its connection to geology. In fact, unlike in *The Child's Botany*, religious references abound. In the early pages of *The Child's Geology* it is clearly Phelps' authorial rather than editorial voice speaking to us when we read "The object of this book is to tell children,- *what the world is made of.* If you ask,- *how was the world made?*—we answer,- *by the Almighty power of God* (1832: 8; emphasis original). She then reminds her audience "You have read in the first chapter of Genesis, that, 'in the beginning God created the heavens and the earth.' We are there told, that 'the earth was without form, and void;' all was in confusion" (Phelps 1832: 8). As she did in her geological lecture for the students of the Troy Female Seminary, Phelps then reviews the events of the six days of creation, and informs her reader that "It is very useful and interesting both for children and grown people, to know something about this world, which God has created on purpose for us, and which is so well fitted for our comfort and for our wants" (Phelps 1832: 8). Note that the language is at a higher level than in *The Child's Botany*, and that she makes specific reference to adults as needing this lesson as much as children.

Phelps' conviction that geology demonstrates irrefutable evidence for the Deluge is also evident, when she writes.

> Geologists are generally agreed upon the fact, that, in former ages, there has been a universal deluge, or flood, which covered the whole earth. They have proved, by examining the surface, that the waves of the ocean have rolled over the tops of the highest mountains and hills. We read of this deluge in the seventh and eighth chapters of Genesis. The appearance of things on the earth, at the present time, proves the account there given to be *true*. (1832: 10; emphasis original)

Finally, Phelps also retreads another of her related lessons from her Saturday geology lecture to the girls at Troy, the important message that the deluge was God's response to humanity's wickedness. In speaking of the diluvial deposits and the fossils found therein she explains.

> The Biblical account of the deluge informs us, that it destroyed all living creatures. Nothing, therefore, can be more evident than that the vast quantities of loose earth, which cover the remains of such myriads of destroyed animals, were deposited by that deluge which God sent, as a punishment to men, to spread desolation over the surface of the earth. (1832: 60)

The text's concluding paragraph is so foreign to both the style and substance of *The Child's Botany* that there is no doubt of the authorship:

> We cannot at this time carry you farther in geological inquiries; after you have understood what is taught in this book, you will be prepared to understand larger works on the same science. The subjects of Geology lead us to contemplate the almighty power and wonderful goodness of God. This power and this goodness are manifested by his operations upon the earth, and by the provisions he has made for our comfort and happiness. When therefore we view the wonderful operations of Nature, we should remember Him, who reigneth over all things; and as our knowledge of his works increases, we should adore Him the more, and feel the greater gratitude for his kindness. (Phelps 1832: 126)

It is therefore apparent that, no matter how many pages of text Phelps had initially received written in the original author's hand, after her pen had been wielded in bold, broad strokes what remained of the voice of the author of *The Child's Botany* was spoken only in a whisper.

This being true, whatever praise or reprobation was meted out by the critics are hers to accept. *The New-York Mirror* called the book "an appropriate guide for young pupils. It contains most valuable information, intelligible to youth" (1832: 167), while a reviewer from the *American Annals of Education and Instruction* was more conflicted over this text, writing:

> We are pleased with this as an introduction to Geology. We cannot, however, consider it as having a right in the title of "The Child's Geology;" for we think, there is either too much of the science or too little of explanation for this purpose. For the higher classes of schools, already familiar with Geography, it may be made very interesting and useful; and like every work of this kind should be illustrated by a collection of specimens. The spirit of the work is excellent; some of the digressions seem to us of doubtful utility. (1833: 191)

It remains unclear whether or not Phelps' heavy-handed discussions of the Deluge and its presumed connection to the geological record are among the "digressions" referenced here. Given the fact that Phelps is directing her reader's attention away from her hand in her preface, instead lauding the deceased originator of the book's concept, and as seen through the eyes of a reader who is unfamiliar with Phelps'style it would be hard to recognize that it was largely her voice in his or her head. Therefore it cannot be maintained that there is an unambiguous female authority figure speaking to the reader, and hence this work does not qualify for inclusion in the familiar format. It is, however, a fascinating example of how a revised work can portray a far different message than the original, more so than in the case of Jane Marcet's *Conversations for Children* and its unauthorized American revision (Sect. 3.3).

8.4 The Politics of Textbooks

In the preface to her *Outlines of Geology for Families and Schools*, Rosina Zornlin notes of geology that "Some acquaintance with its principles... forms an essential part not only of what is usually termed a liberal education, but may also be advantageously extended to all classes of the community, to the cotter no less than the large landed proprietor, to the artisan as well as the master manufacturer, to the miner as well as the owner of mines" (1852: vii). The title of her work was very intentional. Recall that the initial market for children's books was a family market, for leisure time reading as well as for the use of mothers homeschooling their children. In America, the development of the public school system as well as private seminaries for boys and girls opened up a market for standard textbooks. Over 1000 seminaries were founded in America by 1830, the number quadrupling from that in a mere few decades (Rudolph 1996). There was moreover specifically a need for American textbooks that used examples from American society and natural history to illustrate both scientific principles and moral lessons (Boas 1971). For example, Jane Marcet's

Conversations on Chemistry was the most successful elementary chemistry text in America in the first half of the nineteenth century, with American printers alone issuing 23 editions (Lindee 1991). Recall that American publishers also had no compunction about adapting the works of European authors such as Marcet for their own audience, but this took the time and effort of a knowledgeable editor.

To fulfill this increasing need, American scientists and professors took to writing textbooks, both for their own use and to sell to other colleges as a source of income. At Yale, Benjamin Silliman wrote successful textbooks in chemistry and physics, while Elias Loomis wrote over 20 books in mathematics and various topics in natural philosophy and natural history over a 30 year career in the mid-1800s, selling over 500,000 copies in total (Lucier 2009). But many American teachers were not only interested in the scientific content of their courses, but the pedagogy as well. For example, this had been the reason for Emma Willard and William Woodbridge's parallel development of novel geography curricula. Such a shift in the teaching of science was especially true in the case of Amos Eaton, and the female teachers who studied under him, such as Almira Phelps. Eaton's science pedagogy was hands-on and laboratory based, and utilized field trips, student seminars and presentations rather than the increasingly popular lecture and rote memorization (Rudolph 1996).

This new style of secondary education required the development of a new type of textbook. At the time, teachers could choose to use comprehensive, information-heavy textbooks that were written in an impersonal, technical third-person style and left little room for originality or personality. These books were not, either then or now, the source of much enthusiasm on the side of the students. The other choice was popular-level works, many written in a conversational style, some in a fully developed familiar format. These works typically did not cover scientific material in sufficient quantity or depth to be used in schools at anything but the most introductory level. In response to this less than satisfactory situation, Almira Lincoln Phelps began turning her lecture notes from her classes at the Troy Female Seminary into textbooks, incorporating within the texts information on the hands-on activities she included in her own classes, including descriptions and illustrations of simple experiments and observation exercises, such as dissecting flowers (Bolzau 1936). The first of these works was her *Familiar Lectures on Botany* (1829), written with the advice of her former teacher, Eaton.

In her introduction to this work, Phelps[4] noted that "The study of Botany seems peculiarly adapted to females; the objects of its investigation are beautiful and delicate; its pursuits, leading to exercise in the open air, are conducive to health and cheerfulness" (Lincoln 1836: 14). The study of this science was seen to develop both the mind and body of the young girl who studies it. But as Phelps believed there is a decidedly spiritual aspect to the study of nature, the study of Botany also develops the soul. For example, she argues that "The universe consists of *matter* and *mind*. By the faculties of mind with which God has endowed us, we are able to

[4] The convention of referring to her as "Phelps" despite the fact that in 1829 she had not yet married her second husband continues from previous sections. The work was published under the name Lincoln, even in some later editions.

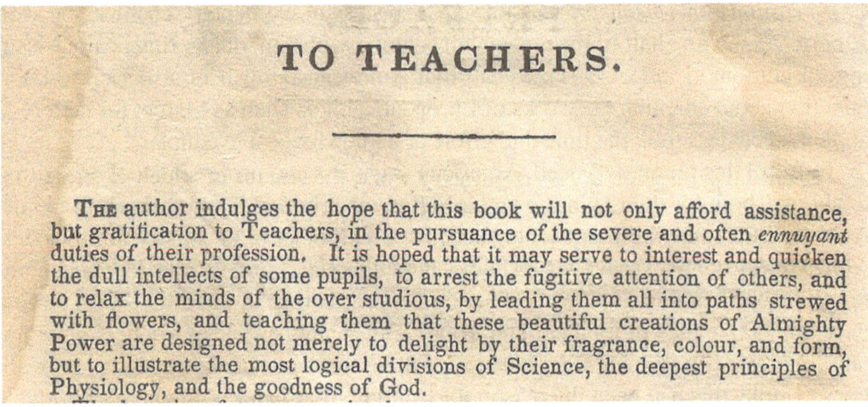

TO TEACHERS.

The author indulges the hope that this book will not only afford assistance, but gratification to Teachers, in the pursuance of the severe and often *ennuyant* duties of their profession. It is hoped that it may serve to interest and quicken the dull intellects of some pupils, to arrest the fugitive attention of others, and to relax the minds of the over studious, by leading them all into paths strewed with flowers, and teaching them that these beautiful creations of Almighty Power are designed not merely to delight by their fragrance, colour, and form, but to illustrate the most logical divisions of Science, the deepest principles of Physiology, and the goodness of God.

Fig. 8.2 Directions to botany teachers. Reproduced from Almira H. Lincoln (1836) *Familiar Lessons on Botany*

examine into the properties of the material objects by which we are surrounded" (Lincoln 1836: 13).

As this book was meant to be used by teachers in the formal classroom, Phelps includes a special set of instructions to teachers at the start of the book (Fig. 8.2), giving tips on what she considers proper pedagogy. For example,

> On the first meeting of the class, after some explanation as to the nature of the study they are about to commence, each member is presented with a flower for analysis. The flower selected is always a simple one, exhibiting in a conspicuous manner the different organs of fructification; the lily and the tulip are both very proper for this purpose. (Lincoln 1836: 8)

She then suggests that the names of the different parts of the flower be explained contemporaneous to the student examining these individual parts through dissection. In this way the students learn the vocabulary in a relational way, while engaging in the process of identification. She also explains that the goal of proper instruction is to bring the student to an appreciation that "these beautiful creations of Almighty Power are designed not merely to delight by their fragrance, colour, and form, but to illustrate the most logical divisions of Science, the deepest principles of Physiology, and the goodness of God" (Lincoln 1836: 8).

Christine Marie Hilger (2008) argues that although *Familiar Lectures on Botany* is not written as a two-way dialogue, it is conversational. The author's tone is less formal (addressing the student directly as *you* and occasionally offering an inclusive *we*) and is decidedly encouraging. Phelps asserts herself as an authority in the subject but is not authoritarian in the way that she does it. Her style has a recognizably maternal edge to it, as she plays the role of supportive mentor helping the student come to a mastery of the content. In this way we can understand how this work can be seen to fit in the continuum of styles included within the familiar format. An example of this supportive, maternal ethos is found in Lecture 3. Phelps warns "When you begin to analyze plants, you will meet with many new terms. It will be necessary in these cases, to resort to the vocabulary of botanical words; by the observation of plants, connected with definitions, you will soon become familiar with the technical terms of Botany" (Lincoln 1836: 19). She then carefully leads the students through the process of analyzing a flower in order to correctly identify the plant. Her directions parallel the pedagogy she has already explained to the teacher in the introduction, and is assuming that her instructions will be followed in the classroom. Also in parallel to the instructions for teachers, she relates her theological intentions to the student early on in the book. She explains that young persons should study botany because.

> the *heavenly bodies* are far distant from us; and were they within our reach, are too mighty for us to grasp; our feeble minds seem overwhelmed in the contemplation of their immensity. *Animals*, though affording the most striking marks of designing wisdom, cannot be dissected and examined without painful emotions. But the *vegetable world* offers a boundless field of inquiry, which may be explored with the most pure and delightful emotions. Here the Almighty seems to manifest himself to us, with less of that dazzling sublimity which it is almost painful to behold in His more magnificent creations; and it would seem, that accommodating the vegetable world to our capacities of observation, He had especially designed it for our investigation and amusement, as well as our sustenance and comfort. The study of Botany naturally leads to greater love and reverence for the Deity. (Lincoln 1836: 15; emphasis original)

Note the reference to design, hearkening back to the natural theology of William Paley (Sect. 4.3).

Phelps' book was extremely successful, and was adopted by many female seminaries. During the four decades that it remained in print it went through two major revisions and nearly 40 printings, and sold well over 200,000 copies and perhaps nearly twice that many (Rudolph 1984). But while Phelps' textbook was an incontrovertible success, it was not the *most* popular work of its type. Harvard professor Asa Gray, author of the more successful and authoritative *Elements of Botany* (1836), criticized Phelps' use of the Linnaean system of classification, which did not allow for natural variations of individuals within a species. As a professional scientist (and beneficiary of formal scientific education), he felt that the work of the largely self-taught Phelps was amateurish (Hilger 2008).

Phelps also published a simplified version of *Familiar Lectures on Botany* entitled *Botany for Beginners: An Introduction to Mrs. Lincoln's Lectures on Botany* (1833), which was also a commercial success, remaining in print for decades with over 20 printings (Rudolph 1984). According to the publisher, the first edition sold out in less than 6 months (Phelps 1849). The work also includes a note to teachers, explaining that the intended audience is students in primary schools and younger students in general. She also offers it as background to any teacher who lacks formal training in botany but "who is in any degree fit to be such" a way to introduce themselves to the topic sufficiently to teach it (Phelps 1849: 8). The book also stresses the use of the senses to observe plants, but adds that "Your *emotions*, too, will be warmed by the thought of His love and kindness who causeth the earth to bring forth, not only 'grass for the beasts of the field, and food for the use of man,' but a rich succession of curious and lovely blossoms for our admiration and enjoyment" (Phelps 1849: 9–10). Phelps also adapted her lectures on natural philosophy and chemistry into two books of *Familiar Lectures* on those topics, as well as the accompanying simplified books for beginners. These four volumes never reached the popularity of the works on botany, despite the fact that they were written in similar styles. It has been argued that at least in the case of *Familiar Lectures on Chemistry* (1838) this was precisely the reason, in particular her heavy-handed religiosity (Lindee 1991).

As seen in the work of previously described women, examples from botany, anatomy, astronomy, and the geology were often used as evidence of the benevolence and wisdom of the deity and, in many cases, a Paley-style argument of design in the natural world. But such arguments had become more subdued in the case of natural philosophy (physics) since the time of Margaret Bryan in the previous century, and had played a far less important role in the dominant familiar format text, Marcet's *Conversations on Chemistry*. Phelps was clearly an exception to this convention. For example, the preface to Phelps' *Familiar Lectures on Natural Philosophy* (1837) begins with a standard explanation that she

> has adopted the style of familiar address, because, while equally favourable for the communication of knowledge, it is more interesting to the pupil than a formal didactic style. A main feature in her manner of treating the subject, consists in bringing forward facts of common occurrence, to illustrate principles; thus teaching the young to reason for themselves. (v)

Where she stands out is the following bluntness of her theological admonition to "American parents and teachers" that her volume is meant to not only "impart to your children and pupils, the principles of science, it will at the same time exert a salutary influence on their moral and religious affections" (Phelps 1837: v). She continues to explain that

> If God is the *eternal geometrician* who has calculated with exactitude the measure of the different forces in the universe, if he is the *wise legislator* who has engraven his laws upon our souls, if he is the *supreme artist,* who has spread forth beauty upon the earth and in the heavens, and has rendered us sensible to its charms; and as there is nothing in the physical world which is not the immediate work of God, and nothing in the moral world, which does not result from faculties with which he has endowed man, there can be neither object nor thought, which may not be connected with God. (Phelps 1837: vi: emphasis original)

This tone of overt natural theology also appears in *Natural Philosophy for Beginners: Designed for Common Schools and Families* (1840). In the preface to this work she offers a quite standard refrain that "The attention of the young should be directed to natural operations; that thus the powers of observation and comparison may be developed and strengthened," but then goes further, arguing that the ultimate goal of such an education is "for inducing in the young mind the grand truth... that, for every thing there must be a cause, until we ascend to the *one great, First cause*" (vi–vii). In order to understand God, she offers, it is necessary to first study his natural laws. As she explains within the text itself, the study of natural philosophy "leads to useful inventions; to the habit of observing and reasoning; it elevates and enlarges the mind, and gives the soul new and delightful views of the power and goodness of God" (Phelps 1840: 13).

Familiar Lectures on Chemistry (1838) has a different history than its related volumes. Due to a lack of time, Phelps had given her lectures notes on culinary and household chemistry to Professor William F. Hopkins (1802–1859) of the United States Military Academy at West Point, not far from Troy, in order for him to assemble them into a volume for her. Hopkins never completed the task and Phelps published it herself in 1838 (Weeks and Dains 1937). Like *Familiar Lectures on Natural*

Philosophy, this volume was only issued twice. A simplified version, *Chemistry for Beginners*, appeared in 1834 and was last issued in 1866, and wore its religiosity on its sleeve. For example, she explains to her young audience that "Wherever you look around, *you behold God's workmanship*; and should you not like to know, *what the objects in nature are made of, and how they are put together*? This knowledge is called Chemistry" (1834: 8; emphasis original). This work also contained a clear moral lesson specifically for the "Young ladies who attend this study," directing them to "pay strict attention to all those facts in housekeeping which may be explained upon chemical principles, such as the action of yeast upon flour, and of pearlash upon sour dough, the change of cider into vinegar… they should, in short, endeavour to make a practical application of their knowledge to the business of life" (1834: 11–12). In addition, they are admonished to remember that

> there is a portion of ourselves which is beyond the scope of chemical science, which cannot be analyzed, because it is incapable of being separated into parts. It is that within us which thinks and feels, which knows good from evil, which is destined to an immortal existence, and which at death passes from its prison of clay to the world of spirits. All then that is done for the improvement of the mind either in knowledge or virtue, will be permanent, while the labour bestowed upon the care and decoration of the body will perish with that frail and decaying substance. (1834: 12–13)

We now return to the first of her mysterious works, mentioned in passing as a footnote in her geology lecture contained within her *Lectures to Young Ladies* (1833: 236):

> Since this lecture was delivered, the author has been engaged in making some additions to a small work written by the author of the Child's Botany, and entitled the *Child's Geology*. This will soon be given to the public. Should the author of these lectures be enabled to fulfil [sic] her present intentions, a work on Geology, adapted to female seminaries, for which she has six years been collecting materials, will soon be prepared for publication.

Henry Barnard refers to a work entitled "Geology for Beginners," which he describes as "a small volume which she never had the courage to revise, as the science has so shaken off its old distinctions, and become essentially changed by the force of new discoveries" (Barnard 1868: 615). Given the other volumes with similar names that she had published, such a volume would have seemed to have been natural for her to produce. However, it seems likely that Barnard was confusing "Geology for Beginners" with *The Child's Geology*. This makes sense, as *The Child's Geology* appears to have never been revised.

But that still leaves open the case of the aforementioned work on geology "adapted to female seminaries." This presumably would have been entitled *Familiar Lectures on Geology* (or perhaps Natural History). Six years is a not insignificant amount of time to work on such a project with no tangible outcome. Perhaps she decided that, as in the case of a potential revision of *The Child's Geology*, the changes in the geological sciences during those decades had become overwhelming. Or, perhaps, it was only one particular type of change that was proving problematic, the slowly changing tide against the Diluvianists involving the realization that the diluvial deposits were not worldwide (a vindication of Cuvier's original writings) and that they were, in fact, not evidence of floods but the actions of ice. Recall that Edward Hitchcock had struggled with these revolutions in geology, punctuated by his own observations of the New England strata, as noted in his writings during the late 1830s and beyond (Sect. 1.3). Perhaps Almira Phelps had not been able to come to an acceptable intellectual compromise, as Hitchcock had done, and simply decided not to complete her book.

While the overt religiosity of Phelps' textbooks appears shocking today, and was considered at the very least old fashioned if not unprofessional by some of her contemporaries, its message clearly resonated with a not insignificant number of readers. In addition, as her goal was the education of her students, and she clearly felt that they were best served by works that spoke directly to what society expected of them (including the roles of Christian wife and mother). She was therefore far more concerned with what young ladies (and presumably their parents) felt about her lessons than the scientific community. As she explained in *Chemistry for Beginners*,

> A scientific professor once said to a lady, in a sarcastic tone – 'and so *you* are going to attempt to teach Chemistry.' 'Yes,' said she, good naturedly, 'as fast as I can learn it myself.' The same professor was, afterwards, engaged to lecture and perform experiments before this lady's class, but her pupils complained that his language was so little adapted to their attainments, that they could neither understand his lectures or experiments, until explained to them in the familiar language of their teacher. (Phelps 1834: iii)

Even when she was not directly in the classroom, Almira Hart Lincoln Phelps was student-centered in her approach, a philosophy that can resonate with professors of all genders who teach introductory or general education courses today.

References

American Annals of Education and Instruction (1833) The child's geology. 3:191
Barnard H (1868) Educational biography. Mrs Almira Lincoln Phelps. American Journal of Education 17(4):611–622
Boas LS (1971) Women's education begins: the rise of the women's colleges. Arno Press, New York
Bolzau EL (1936) Almira Hart Lincoln Phelps: her life and works. The Science Press, Lancaster
Calhoun D (1984) Eyes for the Jacksonian world: William C. Woodbridge and Emma Willard. Journal of the Early Republic 4(1):1–26
Goodrich SG (1831) The child's botany, 4th edn. Carter and Hendee, Boston
Goodrich SG (1857) Recollections of a lifetime, or men and things I have seen, vol 2. Miller, Orton and Mulligan, New York
Grigg S (1999) Willard, Emma Hart. In: Garraty JA, Carnes MC (eds) American national biography, vol 23. Oxford UP, New York, pp 408–410
Hilger CM (2008) Almira Hart Lincoln Phelps. In: Patterson D (ed) Early American nature writers: a biographical encyclopedia. Greenwood, Westport, pp 281–287
Lincoln AH (1836) Familiar lectures on botany, 5th edn. Huntington, Hartford
Lindee MS (1991) The American career of Jane Marcet's Conversations on chemistry, 1806–1853. Isis 82(1):8–23
Lucier P (2009) The profesional and the scientist in ninteenth-century America. Isis 100(4):699–732
Phelps AHL (1832) The child's geology, rev. Peck, Brattleboro
Phelps AHL (1833) Lectures to young ladies. Carter, Hendee and Co., Boston
Phelps AHL (1834) Chemistry for beginners. Huntington, Hartford
Phelps AHL (1837) Familiar lectures on natural philosophy. Huntington, New York

Phelps AHL (1840) Natural philosophy for beginners: designed for common schools and families. Huntington, New York
Phelps AHL (1849) Botany for beginners. Huntington and Savage, New York
Rudolph ED (1984) Almira Hart Lincoln Phelps (1793-1884) and the spread of botany in nineteenth century America. Am J Bot 71(8):1161–1167
Rudolph ED (1996) History of the botanical teaching laboratory in the United States. Am J Bot 83(5):661–671
The New-York Mirror (1832) Our book-table. 10(21):167
Weeks ME, Dains FB (1937) Mrs. A.H. Lincoln Phelps and her services to chemical education. J Chem Educ 14(2):53–57
Woodbridge WC, Willard E (1824) Universal geography, ancient and modern. Cooke and sons, Hartford
Zornlin RM (1852) Outlines of geology for families and schools. Parker, London

Chapter 9
Alternatives to the Familiar Format

9.1 Changing Audiences

The familiar format was one of a number of different literary styles adopted by popularizers of science in the first half of the nineteenth century. Many authors, especially women, felt comfortable utilizing it, and felt it well served their intended audience of children and potential teachers (including mothers). But there were other audiences to reach, as well as authors who were not predisposed to writing in a conversational style. This chapter surveys examples of both of these.

It is important to note that one of the most successful popular-level works of geology in the mid 1800s was not written in the familiar format nor addressed to children. Hugh Miller's *The Old Red Sandstone or New Walks in an Old Field* (1841) was published in least 25 different printings (Lightman 2007). The editor of *The Witness* newspaper and later a member of the Free Church of Scotland, Miller originally published a third of the reminiscences, vignettes, and pieces of advice and encouragement related to both ethics and education that make up the book as a "series of sketches in the *Witness* newspaper. A portion of the first chapter was submitted to the public a year or two earlier, in *Chambers' Edinburgh Journal*" (Miller 1841: vii). In his preface he apologizes to his two intended audiences, geologists and general readers, noting that he could not satisfy both simultaneously, as his work included both "familiar phenomenon" and "descriptions of hitherto undescribed organisms, and the deposits of little-known localities" (Miller 1841: viii).

But there is a very specific audience to which Miller addresses his opening remarks, "young working men desirous to better their circumstances, and add to the amount of their enjoyment." He specifically encourages them to

> not seek happiness in what is misnamed pleasure; seek it rather in what is termed study. Keep your conscience clear, your curiosity fresh, and embrace every opportunity of cultivating your minds.... Learn to make a right use of your eyes; the commonest things are worth looking at – even stones and weeds, and the most familiar animals. Read good books, not forgetting the best of all: there is more true philosophy in the Bible than in every work of every sceptic [sic] that ever wrote.... (Miller 1841: 1–2)

K. Larsen, *The Women Who Popularized Geology in the 19th Century*,
DOI 10.1007/978-3-319-64952-8_9

Miller also warns against wasting time being jealous of the upper class, instead admonishing his readers that "there is only one way in which your jealousy of them can be well directed. Do not let them get ahead of you in intelligence" (1841: 2). He then begins to tell the tale of his first day at work in a quarry, hard, dirty work that was nonetheless fascinating as it allowed him to examine the strata for himself, and afforded him opportunities to increase his knowledge of geology firsthand. In Miller's work we witness a turning away from works for children, even older children, or women, and specifically toward working class men who would seek to better themselves through the acquisition of scientific knowledge. While such an audience also played a central role in the Lyceum movement, the writing of a work specifically for their benefit and needs was a significant shift in the popularization of science.

Miller's work is just one example of a geological popularization that was not intended for the stereotypical audience that the familiar format had presumably served so well. Other works of this same era, written by women, also deviated from this, by now traditionally feminine, genre, and demonstrated that women writers were willing to push the envelope of science popularization in many different directions.

9.2 Mary Roberts (1788–1864): Young Earth Literalist

A recurring theme in this volume has been the continuum of relationships depicted between religion and geology, with the majority of geologists and popularizers of this time occupying a space between strict young earth Scriptural literalism and a complete separation between science and theology. Mary Roberts (1788–1864) is a fascinating exception to this trend. Born in London on March 18, 1788, Mary was the only daughter of four children born to American merchant Daniel Roberts (1753–1811) and his wife Ann Thompson. She was brought up in a Quaker household, and one of her grandfathers was Thomas Lawson (1630–1691), a Quaker botanist (Fell-Smith 1896). Mary Roberts' first encounter with geology was not a particularly happy one. In 1790 her father became involved in an ill-advised attempt to find coal in Gloucestershire and became a follower of the self-described prophet Joanna Southcott (1750–1814). Southcott convinced her followers that she was the Woman of the Apocalypse foretold in the Book of Revelations, and that the seals that she sold offered not only the true wisdom of God, but protection from Napoleon and a guarantee that they would reign with Christ on earth (Denham 1815). Daniel Roberts was expelled from the Society of Friends in 1807 for his support of Southcott, and his family resigned as well in protest. After her father's death in 1811 Mary returned to London with her mother. She never married. One of her brothers, Oade (1786–1821), returned to the Society in 1813, and Mary, along with her mother and another brother, John, did the same in 1818 (Urdank 1990).

While coal was not found in Painswick, Mary found an interest in the countryside and its natural beauty, as did Oade, who became an amateur botanist. During these years she wrote the work for which she was later most famous, *Annals of My Village*. A month-by-month description of her observations of both the environment and life in a small village, the work was not published until 1831. She is also known for many less famous works, including *Select Female Biography* (1821). She also wrote a number of popular-level books on natural history, including *The Wonders of the Vegetable Kingdom Displayed* (1822), *The Conchologist's Companion* (1824), *The Seaside Companion, or Marine Natural History* (1835), *Sketches of the Animal and Vegetable Productions of America* (1839), *Voices from the Woodlands, or History of Forest Trees, Ferns, Mosses, and Lichens* (1850), and *A Popular History of the Mollusca* (1851). She also produced *Sister Mary's Tales in Natural History* (1834), which earned her the nickname of Sister Mary (Shteir 1996).

All of her books reflect the same unwavering fundamentalist faith but varied in their literary style (Leach 2006). For example, *The Wonders of the Vegetable Kingdom Displayed* is in the form of letters to an unnamed person, but as Ann Shteir has argued it varies from the familiar format in many important regards, including its lack of a mother figure who imparts knowledge or familial setting. Instead, the letters are written by characters with decidedly mythological sounding names, such as Orontes and Laelius (Shteir 1996). The advertisement page of this work explains that "The Author's aim is merely to profit and to please; to lead the youthful mind from the minor wonders of creation to the knowledge of their great Artificer; and the admirers of the elegant and captivating science of Botany to consider how much it is capable of being heightened and improved by associations of the highest nature" (Roberts 1824: np). Paley's Artificer God is repeatedly referenced within the work, where he is termed "the great Author of the universe… the Divine Artificer" who designed each plant and animal "perfect in its kind" (Roberts 1824: 238–39).

While paleontologist Stephen Jay Gould (1997) notes that Roberts' natural history works generally conform to the scientific standards of her time, her interpretation of the geological record, published as *The Progress of Creation* (1837), was another matter. As she explains in her preface, "Throughout this volume, I have ever kept in view, that the heavens, and earth were finished, and all the host of them in six days; and that no theory, however plausible, can be admitted in opposition to the Divine Record" (Roberts 1838: vi). She specifically proclaims that

> We learn from the Mosaic record, that the first act of creative power gave existence to the mighty nucleus of our globe, and that, before the revolution of seven days, the herbless granite was clothed with vegetation; moving creatures then passed through the waters, and winged fowl were seen to fly across the heavens; cattle, and creeping things, and beasts of the earth, walked forth in all the vigour of their new existence, and, lastly, man, created, in the image of his Maker, had dominion over every living thing. (Roberts 1838: v)

She utterly rejects attempts to reconcile geology and theology through a figurative, metaphorical interpretation of Scripture. Instead she agrees with Bacon and Newton, who

> received the words of scripture as Moses intended them. They believed that each day was such as is understood by the common acceptation of the word; a natural day; the revolution of twenty-four hours, commencing from one evening, and terminated by the next, according to the reckoning of the most ancient nations. (Roberts 1838: 4)

In keeping with her literalist interpretation, the book is laid out with one chapter representing each day of creation, covering the six days of creation plus the present day.

Each chapter begins with a quotation from the relevant portion of Genesis (referring to that day of creation), and whenever the Bible seems contradicted by modern science, she weaves an explanation to bring the two descriptions into harmony. For example, in her discussion of the first day she explains that earth was actually hidden or shrouded until God made the decision to reveal it, citing earlier authors such as Josephus.

It is not surprising that Roberts feels she "must confess my obligations to the learned author of the Comparative Estimate of the Mineral and Mosaical Geologies [of Granville Penn]," (1838:vi), whose interpretation of the geological record she largely adopts, albeit with the important caveat that Penn was not a young earth literalist (Sect. 9.2). She also credits Cuvier's *Essay on the Theory of the Earth*, a work which, as previously explained, was not intended by its author to support a literalist interpretation of the Bible, although often misused by Diluvianists as such (Sect. 9.2) In particular, Roberts cites Cuvier's work on fossil shells as unambiguous evidence of Noah's flood:

> Every thing concurs to indicate that this earth has undergone the most important revolutions; sea shells and the bones of land animals encrusted in mountains, clearly prove this most extraordinary fact .... Every portion of the habitable globe exhibits the same phenomena, and we are therefore led to the conclusion, not only that the sea at one period covered the surface of the earth, but that it must have remained there for a long time, and in a state of tranquility, which circumstance was necessary for the formation of deposits so extensive, so thick and solid, and containing exuviae so perfectly preserved. (1838: 228–9)

She further interprets this to mean that the evidence of geology supports "two great convulsions of the earth's surface," the first referring to the Biblical separation of land from the water (before the existence of life) and the later Noachian cataclysm. In her words,

> The first, occasioning through the violent action of mechanical agencies, directed by the Creator, a bed to receive the waters, which had previously been diffused over the whole earth; the second, producing under the same direction, and by a repetition of the same operation, a new bason [sic] into which the waters were transferred from their former bed. It also establishes the fact, of the sea having occupied that former bed, during the lapse of time, that intervened between the creation and the deluge; and finally, of that former bed, being now the earth which we inhabit. (1838: 232)

It should be noted that, although Roberts cites the standard authorities of her day, she does not feel the need to slavishly accept all of their interpretations of the evidence. For example, she takes issue with Cuvier's assumption that the mastodon was an herbivore. Calling it the "carnivorous elephant," Roberts instead describes its "enormous grinding teeth" as more closely resembling those of a carnivore, and suggests that "there is good reason to believe, that the creature preyed on animal food" (Roberts 1838: 243). This is not the only instance in her book where she offers an original (or at least not widely discussed) interpretation of the evidence.[1] For instance, she has a rather creative explanation for the apparent lack of human

[1] Roberts' willingness to offer original interpretations and question the authorities of her day gives Ralph O'Connor (2007) reason to disagree with Gould's (1997) assessment of Roberts as passively accepting oppression by the male authorities of her time.

fossils found mixed with animal fossils in the rock layers attributed to the Flood: due to their lack of intelligence, animals would tend to huddle together in fear at the approaching floodwaters, unable to formulate an escape plan. Therefore animals

> would be carried off in great numbers by the inundation. The human population, on the contrary, being endowed with observation, and strongly actuated by hope and fear, would anxiously retreat before the advancing waters, most probably to the highest point of land: Until at length, being assembled on that narrow spot of earth, which alone remained to them, amid the increasing inundation, they would not have been washed by degrees into the water, and carried away by any reflux, like the poor affrighted animals that crowded together; but absorbed into one fearful vortex, when the last portion of firm land sunk beneath their feet. They would then suddenly be submerged in the depth of the new sea, where their bones must remain for ever beyond the reach of man. (Roberts 1838: 239–40)

As for why some animals became extinct and were not saved by Noah, Roberts echoes Penn (Sect. 9.2), offering "we have every reason to conclude, that when the Most High made known to Noah, the species, which he designed to preserve, he was pleased to except some from that preservation" (1838: 243). She then engages in a bit of Biblical interpretation of her own, arguing that since Noah was directed to "take OF every living thing of all flesh" that meant that not all species would be included (1838: 245). She further argues that "It is certainly not contrary to the Divine Wisdom that certain species should have become extinct, when they had fulfilled the purpose for which they were created" (1838: 245). However, this begs the question, why were some ferocious carnivores (such as lions and tigers) saved on the ark and not others? She offers that perhaps these creatures were not originally as ferocious as they are today and became such only in response to the effects of the hot climates in which they now find themselves. Alternatively, "when the Most High swept from the surface of the earth those terrible creatures, the carnivorous elephant and reptile, with others of a similar description, that he might have replaced them by the present comparatively innoxious race of predactory animals" (1838: 247). Such a special creation of new types of animals is logical to Roberts, as it was to Penn, as

> a new vegetable creation must have clothed the bed of the ancient sea, when it was forsaken by the waters; and why should not the same creative power have also produced animals? In concluding, therefore, either that carnivorous propensities were given to, or drawn forth from certain species; or even that new species were created for a specific purpose, we conjecture nothing at variance with Holy Writ, nor with the deductions of reason. (247–8)

The 1830s was a decade in which Diluvianists were faced with an ever accumulating preponderance of evidence that the geological record did not, in fact, provide direct evidence for the Deluge. Edward Hitchcock had struggled with this reality, and was able to achieve a reconciliation that personally satisfied both his beliefs as a Christian and his work as a geologist. Mary Roberts represents a very different type of resolution, in which the book of Scripture was clearly seen to trump the book of strata, and the efforts of the limited human mind to understand God's creation. As she warns,

> Let us also ever bear in mind that the science of geology is in its infancy, that it has not yet emerged from the dim twilight of past years, that man has penetrated but a little way into the strata of the earth. All that we really know respecting the creation of the world, and the

> era of the flood is derived from the record which the Most High has given to his creatures. This must not be impugned at will. It is too sacred to be lightly canvassed. Where the discoveries of modern date, do not harmonize with the facts therein recorded, it becomes us to be silent, and to confess, that we know nothing. (1838: 251–2)

This was an admonition that the geologists of the nineteenth century were certainly not willing to obey.

9.3 Rosina Zornlin (1795–1859): Adapting to the Audience

While Jane Marcet embraced the familiar format for the majority of her works, there were women such as Delvalle Lowry and Maria Hack who adapted their literary style to the individual work and its intended audience. An example of a woman who produced some works reminiscent of the familiar format but elected to avoid a conversational form for her works in the geological sciences was Rosina Zornlin (1795–1859). The middle daughter of London businessman John Jacob Zornlin and his wife Elizabeth Alsager, Rosina was born on December 6, 1795 in Walthamstow, Essex, but spent much of her life as an invalid residing in the family's home in Clapham, Surrey (Creese and Creese 2004). Her mother wrote poetry and had individual pieces published in magazines, and her younger sister Georgiana Margaritta Zornlin (1800–1885?), who was interested in archaeology, published at least three short works: *A Paper Lantern for Paseyites* (1842), *Urim and Thummim, an inquiry* (1860), and *The Heraldry of the World, Observations on the Universality and Antiquity of the Seal* (1874). The last of these was a paper that had been read at the Winchester and Hampshire Scientific and Literary Society. Georgiana also wrote letters that were published in the journal *Notes and Queries* (Thomas 1905). Rosina died on May 22, 1859 at Kenilworth.

Despite her frail health, Rosina engaged in amateur astronomy and published two articles on meteor showers in the *Philosophical Magazine*, one strictly observational (her observations of what we now call the Perseids [Zornlin 1839b] in response to a published request to submit such observations) and the second a more theoretical piece trying to connect aurora and meteor showers (and other electrical phenomenon), although it did include some observations of what are most likely Leonids (Zornlin 1841). Her interest in physics was also evidenced by an 1858 paper entitled "On Heat and on the Indestructibility of Elementary Bodies" that was read to the British Association for the Advancement of Science by S. W. Ayerton (Zornlin 1859).

Zornlin wrote on both scientific and religious topics. Two nonfiction religious works, *Bible Narrative, chronologically arranged in the words of the authorized version* [of the Bible] (1838) and *Questions on the Bible Narrative* (1844) were published, along with an anti-Catholic novel *The Roman Catholic Chapel, or, Lindenhurst Parish* (1837) in which the Catholic Church is compared to a cat that sinks its claws into its adherents (*The Dublin Review* 1838).[2] Among Zornlin's popularizations of science are three written in dialogue form, but all lacking a female

[2] More information on Zornlin's religious writings can be found in Creese and Creese (2004).

teacher/tutor. *What is a Comet, Papa? or, familiar descriptions of comets; more particularly of Halley's Comet* (1835) was published to coincide with the apparition of that famous comet. The story is constructed in two parts; in the first, a boy leads his sisters in a discussion about the comet, and in the second the father explains what he knows about the celestial visitor. An anonymous reviewer in *The Spectator* (1835: 760) somewhat mockingly notes

> Miss Zornlin probably imagines that the vulgar and childhood may apprehend danger from the approaching comet. She has therefore come forward to tell them, in a very simple and interesting manner, not only the "heads" of what is known about comets, but also to explain the leading principles of the heavenly bodies. We can recommend the little tract, as well adapted for the objects the fair writer appears to have in view.

The Solar Eclipse, or, the Two Almanacks (1836) was published ahead of the annular solar eclipse of May 15, 1836, visible in parts of the UK. The eclipse is now considered to have had particular scientific importance because it marked the first description of a phenomenon now called Baily's Beads by astronomer Francis Baily (1838). A luminous string of beads of light appears at the edge of the moon just before or after totality (or on the limb of the moon in an annular eclipse) due to sunlight shining through craters. The dialogue is between Charles Walcot and his father, and includes both astronomical and theological information. An unnamed reviewer in *The Monthly Review* (1836: 185) lauds both Zornlin's "unassuming graceful taste" and the "simplicity, the attractive and comprehensive character of the lessons conveyed by the dialogue" that are suitable for both children and "uninformed adults.... In short, we may call it one of the best little rudiments to the study of astronomy; for while it teaches, it entices, at the same time that it conveys the truest principles of religious feeling." The reviewer specifically quotes one of the religious lessons of Mr. Walcot as an example: "The more we penetrate into the wonders of Creation, the more shall we perceive the harmony, and the order, and beauty, which pervade the whole system of the universe; and the better be able to comprehend the power and goodness of that God, to whom we owe every blessing here, and every hope of happiness hereafter" (*The Monthly Review* 1836: 186).

The third of the dialogue-based works was previously mentioned in Sect. 6.3, *What is a Voltaic Battery*?, along with its less than enthusiastic review by surgeon Charles Morgan in his diatribe against the genre in general. Of Zornlin's work in particular he notes "the little volume is by no means worse than the greater number of its class; and we have adopted it to illustrate the general fact" (Morgan 1842: 834). The work is written as a conversation between Edward, his brother William, and his uncle, Mr. Compton. As there are long expositions by Mr. C. the conversational form is not particularly well used, and the lack of a female authority figure, perhaps not unexpected given that the topic is physics and not botany or natural history, flags this work as not a true example of the familiar format.

The work begins rather abruptly, with the titular question:

> "What is a voltaic battery?" said Edward Compton to his uncle.
>
> "Why, surely, you know," interrupted his brother William: "do you remember seeing one at the Adelaide Gallery, some time since?"

> "I well remember," replied Edward," having seen some very remarkable experiments performed by means of a voltaic battery at that place, but I did not then, nor do I now, understand the principle on which it acted: and I was going to request my uncle to give me some information on the subject." (Zornlin 1843: 3)

Morgan argues that the successful communicator of knowledge must lead the student or reader through the material step by step, carefully building on what has already been mastered, and each step must be small, clear, and detailed. Herein lays his particular complaint with Zornlin:

> It is evident on the simplest perusal, that the writer unconsciously takes for granted that the pupil is familiar with the phenomena which she undertakes to explain; but this familiarity could not have been attained without having placed the pupil in advance of the elementary matter, which such works profess to teach. Here, indeed, lies the root of the whole error – science, being a matter of fact, cannot be taught by books. The electrical machine, and the voltaic battery, must be seen and used before they can be understood. (Morgan 1842: 834)

Note the parallel with Delvalle Lowry's philosophy concerning the teaching of mineralogy (Sect. 6.2).

In contrast, neither of Zornlin's two books on geology —*Recreations in Geology* (1839) and *Outlines of Geology* (1852)—nor the pair of works on physical geography—*Recreations in Physical Geography* (1840) and *Outlines of Physical Geography* (1851)—are written in dialogue style, nor was her work *The World of Waters, or Recreations in Hydrology* (1843). Instead, they are written in clear textbook style prose. The first, *Recreations in Geology*, has a prolonged introduction, entitled "Preliminary Discourse on the Nature and Advantages of Geology," an excerpt of which appeared in *The Saturday Magazine* (Zornlin 1840). The dialogue is curious for its reliance on a significant number of both long (several pages in length) and short quotations, not all of which are attributed. Such sources include Charles Lyell's "Anniversary Address to the Geological Society" (1837), John Phillips' *A Treatise on Geology* (1837–9), William Whewell's *History of the Inductive Sciences* (1837), and other famous works. A 20 page long discussion on the "Utility of Geology" is largely quoted from what she terms "an eminent geologist" and cites as "*Magazine of Popular Science*" (Zornlin 1839a: lix) although it is also nearly identical with a chapter on the practical applications of geology in Joshua Trimmer's later *Practical Geology and Mineralogy* (1841).

Religion plays a central role in this volume and she connects geology and theology on multiple occasions. In some ways her syncretism of geology and theology are generally reflective of similar works, but on occasion does take the argument a step further. Her work is clearly reminiscent of other narratives of natural theology (such as those of Maria Hack) in her open appreciation of William Paley's work (Sect. 4.3). For example, she notes "A very close analogy is thus found to exist between extinct and recent species, so as to leave no doubt on the mind, that the same harmony of parts and beauty of contrivance which we admire in the living creature, has also characterized the organic world at remote periods" (1839a: xxii). Another reference to Paley's theology can be found in a section that includes a short (attributed) quote from William Buckland's *Geology and Mineralogy Considered with Relation to Natural Theology* (1836):

> We have seen that geology has very practical advantages, and that it is also conducive to the improvement of our minds. Nor does it stop here; it takes a yet more exalted stand; this science is at the same time "the efficient auxiliary and handmaid of religion." Wherever we direct our researches in the natural world, we meet with distinct traces of a controlling Intelligence, and thus the employment of our intellectual faculties leads us to the contemplation of the great First Cause. (Zornlin 1839a: lix)

As is common in such works, Zornlin describes the earth as not only the special creation of a benevolent and omniscient deity, but having been specially designed for the comfort of humanity. However, she specifically appeals to a sense of English national pride and Anglican superiority while doing so:

> Thus has the beneficent Creator provided for us an invaluable supply of fuel [coal], in the growth of these immense and remarkable [prehistoric] forests. And, in truth, had not this useful material been thus abundantly prepared for us, it would have been impossible to have procured a sufficient supply for all the purposes of comfort and utility to which coal is at present applied. Our manufactories, our steam-engines, never could have been furnished with fuel by any other means; nor could they have been brought to the perfection which has raised Great Britain so high in the commercial world; and which, by extending her influence over the globe, has enabled her to spread to far distant lands, the benefits of civilization and religious truth. (Zornlin 1839a: 54)

Zornlin's interpretation of the geological record is markedly different from that of the Diluvianists, and represents an alternative explanation held by some of her generation. In this view, it was expected that the Noachian flood, being of relatively short duration, would not be apparent in the fossil and rock record. Instead, all of Cuvier's evidence for multiple catastrophes, including the diluvium, described a world *before* the existence of man. Since the diluvium itself (interpreted as representing the last catastrophic event) was relatively recent, and human fossils were found only above this layer, it was interpreted as proof that

> man's existence on earth, cannot be referred to an era more ancient than that assigned in the Sacred Volume to the creation of the human species, but, as nearly as can be expected in such investigations, accords remarkably with that date. This is a conclusion at the same time the most clearly proved, though perhaps the least expected from geology; a conclusion, however, peculiarly valuable, as it forms a link between history and the Bible. (Zornlin 1839a: 183)

Zornlin then openly acknowledges and answers the critics of such an interpretation of the extinctions apparent in the strata, similar to arguments by Maria Hack and Jane Kilby Welsh:

> It has been objected, that the Bible gives no account of these changes which so evidently appear to have occurred antecedent to the creation of the human species. But in fact, such explanations would have been foreign to the design of Holy Writ. We find two special objects kept in view in the Old Testament – the one, to turn men from idolatry; the other, to direct their attention to the promised Messiah: - to neither of these points would a description of the revolutions of the globe in any way have tended. For these purposes, it sufficed to declare that God was the Creator of all things. (1839a: 15)

The detailed depiction of attempts at reconciliation between the theology and geology of the mid nineteenth century suggests that Zornlin's work should be of interest to historians of science.

Her second major work in the geological sciences was a companion volume highlighting the appearance of the modern world, *Recreations in Physical Geography* (1840). Zornlin describes the difference between the two disciplines by analogy, offering that "Geology may be compared to Ancient History, around which distance of time and partial obscurity throw the halo of classic interest. PHYSICAL GEOGRAPHY, on the other hand, resembles Modern History, or rather, perhaps, records of daily recurring events – events in which our own personal interests are concerned" (1842: iii). The writing is clear and matter-of-fact, and lacks the long quotes from experts found in *Recreations in Geology*, although unattributed quotations of a shorter variety can still be found scattered throughout her work. Individual chapters begin with a short quote, but they are topical rather than religious (for example, from Shakespeare on the globe), although some short verses within the text are of a religious nature. In general, there is far less invocation of religion in this text, although references to Paley's theology can still be found. For example, she notes that

> the harmony and glory of creation can be fully appreciated only by those who are in some measure acquainted with the laws of nature. The study of these laws possesses at the same time, the advantage of fixing the mind on nature's Ruler; for a law not only necessarily implies a lawgiver, but if these laws are found to be uniformly and universally carried into effect, we are further led to the perception of the continually watchful Providence, by which the natural world is governed and sustained; and, consequently, to that both of the Omnipotence and Omnipresence of the Deity. (Zornlin 1842: 4)

As is often found in popularized works of this era, she ends with a religious entreaty to her reader, reminding that

> the contemplation of so much that is lovely, so much that is excellent, has, when considered with reference to Him from whom it all proceeds, a direct tendency to improve and elevate the character; whilst the habitual acknowledgment of the hand of the Most High, in these His visible works, cannot fail to confirm and strengthen our trust and reliance in Wisdom, Power, and Goodness so infinite. (1842: 402)

Therefore while the format of her works in the geological sciences do not conform to the familiar format, in their central lesson they are quite similar in many respects.

Zornlin later published briefer volumes in both geology and physical geography under the titles "Outlines", which included exercises that could be used by teachers. Given the much shorter length of these volumes, as well as their intention to be used in schools, it is not surprising that theology plays less of a role in *Outlines of Geology for Families and Schools* as compared with *Recreations in Geology*. But as she ends the slender former volume,

> Brief as are the outlines here presented of the features of Geology, even these cannot fail to excite our admiration and delight, whether we regard this science as a key to the treasures of the natural world, or in the higher light of one of the various means which lead us to the contemplation of wisdom, power, and goodness of the beneficent Creator of all things. (Zornlin 1852: 98)

Outlines of Physical Geography for Families and Schools also appeared in an 1856 American edition revised by William L. Gage, former Master of Taunton High School, entitled *Physical Geography for Families and Schools.* Gage particularly appreciates the religious message of Zornlin's work, explaining in his preface,

> While its range is very broad; while we are now lifted to the clouds, and anon carried far into the depths beneath our feet; transported in a moment from ocean to ocean and from shore to shore, initiated into the manifold secrets of nature, and taught on every side the greatness and the wisdom and the love of the Creator and Father of us all, we see the chain which binds together all the sciences and makes them one. (Gage 1856: xiii)

The North American Review writes of Gage's revision "We like this book, and, were there more of it, *we* should like it better." The reviewer observes that the work's "brevity is secured by condensation rather than omission; and a certain measure of dryness is inevitable from the attempt to compress so much matter within so small a space." For this reason, the reviewer doubts its usefulness as a strict textbook, but offers that its "true use would be as a syllabus for recitation, where the teacher is qualified to accompany it by oral explanations and lectures" (1856: 281). Regardless of the reason, Zornlin's works have since passed into relative obscurity, even among historians of science, and are deserving of more detailed analysis.

9.4 Mary Somerville (1780–1872): A Woman of Science

American astronomer Maria Mitchell wrote of Mary Somerville (1780–1872)

> There have been in every age a few women of genius who have become the successful rivals of man in the paths which they have severally chosen. Three instances are of our time. Mrs. Browning is called a poet even by poets; the artists admit that Rosa Bonheur is a painter; and the mathematicians accord to Mary Somerville a high rank among themselves. (1860: 568)

As Mary and Thomas Creese (2004) observe, Mary Somerville's *Physical Geography* (1848) is often erroneously referred to as the first major work of its kind in English, as Zornlin's volume predates it by nearly a decade. However, given the singular fame of Mary Somerville (1780–1872) in her own time, and the relative obscurity of Zornlin in ours, such an oversight should not be surprising. Much has been written elsewhere about Mary Somerville, as an extraordinary woman who found respect while writing science books clearly targeting adults (specifically adults already beyond the most basic level of knowledge). The current analysis will therefore be limited to the aspects of her life that impacted her *Physical Geography* and situate her in relation to the other women discussed in this work.[3]

Mary Fairfax Greig Somerville was born at Jedburgh in Scotland on December 26, 1780, the daughter of Admiral William George Fairfax, a relative of George Washington. As a child she loved to explore nature on her own and, as was typical of her generation, was a collector of objects such as shells and bird eggs. Not only was she home-schooled in the typical feminine skills such as sewing and painting, but uncommonly taught herself geometry from textbooks provided by her brother's tutor. She married a cousin, Samuel Greig, a captain in the Russian navy and Russian Consul in 1804, but was widowed three years later and left with two small children. She had

[3] For more on Somerville's life, see Patterson (1983).

continued her study of mathematics throughout her marriage (but notably without her husband's encouragement), a situation that changed during her second marriage, to another cousin, Dr. William Somerville, in 1812. As she later recalled, "In those early days I had every difficulty to contend with; now, through the kindness and liberal opinions of my husband, I had every encouragement" (Somerville 1874: 95).

Mary began to work on an exposition of Laplace's *Méchanique Céleste*, including her own mathematical proofs where she felt she could be clearer than the original. Its publication in 1831, under the title *The Mechanism of the Heavens*, brought her the respect of the scientific community, but the work was too complex for consumption by the general public. Fortunately her second work, *The Connexion of the Physical Sciences* (1834), went through ten editions, and brought her wider acclaim as well as a broader audience and a civil pension. Her third book, *Physical Geography*, followed in 1848 and continued to be published until 1877, and after William's death in 1860, she slowly worked on her final book, *Molecular and Microscopic Science* (1869). Bernard Lightman (2007) describes in detail how Somerville's works significantly differ from those in the familiar format in both their literary style and intended audience. The differences are so stark that biographer Elizabeth C. Patterson (1969) argues that Somerville was not truly popularizing science but rather clearly and correctly summarizing and explaining the science, all the while utilizing appropriate scientific terminology. Somerville received many awards during her life, including the Victoria Medal of the Royal Geographical Society (1869), and was made an honorary member of the Royal Irish Academy (1834). In 1835 she and astronomer Caroline Herschel were the first women to become honorary members of the Royal Astronomical Society.

In her autobiographical memoirs Mary Somerville recollects in detail the series of events that led to her studies in mineralogy and later geology, acquiring the knowledge that she would put to excellent use in her *Physical Geography*. She recalls that early in her second marriage she

> had heard that Sir James Hall had taken up the subject, but I did not care about it; I am certain that at that time I had never heard the word Geology. I think it was now, on going with Somerville to see the Edinburgh Museum, that I recognised the fossil plants I had seen in the coal limestone on the sands at the Links of Burntisland. Ultimately Geology became a favourite pursuit of ours, but then minerals were the objects of our joint study. (Somerville 1874: 93–4)

William and Mary began enthusiastically constructing their mineral cabinet and "were criticized for extravagance," although Mary admits that "no doubt I had the lion's share of blame" (Somerville 1874: 95). After moving to London in 1816 she "took lessons in mineralogy from Mrs. [Rebekah] Lowry, a Jewess, the wife of an eminent line engraver, who had a large collection of minerals, and in the evening Somerville and I amused ourselves with our own," (Somerville 1874: 106). The couple continued to collect minerals on their travels, including some that they collected on their own, as well as gifts and purchases. For example, Scottish naturalist George Finlayson (1790–1823) "who was with our troops in Ceylon, and who had devoted all his spare time to the study of the natural productions of the country, sent us a valuable collection of crystals of sapphire, ruby, oriental topaz, amethyst, &c." (Somerville 1874: 127-128).

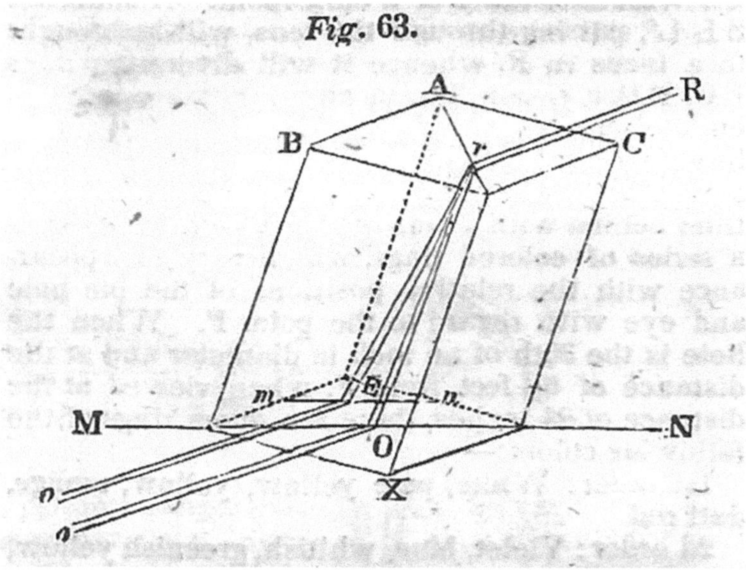

Fig. 9.1 Ray diagram explaining double refraction through a crystal of Icelandic spar. Reproduced from Mary Somerville (1846) *On the Connection of the Physical Sciences*, Amer. edn

The Somervilles analyzed their mineral samples using the technology of the day, often aided by their scientist friends (Fig. 9.1). They not only bought a reflective goniometer, but were taught how to use it by its inventor, their frequent dinner guest William Wollaston (Sect. 9.1), who bequeathed to his friend Mary Somerville "a collection of models of the forms of all the natural crystals then known" (Somerville 1874: 129). But it was William's hobby of using a blowpipe to analyze their samples that nearly proved fatal to his wife. As she later described,

> One evening, when he was so occupied, I was playing the piano, when suddenly I fainted; he was very much startled, as neither I nor any of our family had ever done such a thing. When I recovered, I said it was the smell of garlic that had made me ill. The truth was, the mineral contained arsenic, and I was poisoned for the time by the fumes. (Somerville 1874: 128)

Although perhaps unrelated to this unfortunate incident, Mary Somerville ultimately became more interested in geological formations than minerals, and learned much from geologist friends, including Sir Charles Lyell and Sir Robert and Lady Charlotte Murchison.

In 1838, William, then in ill health, retired from Chelsea Hospital. The couple moved to Italy where the climate was less stressful on William. Here Mary slowly wrote *Physical Geography* (Fig. 9.2). Upon its completion she learned that Alexander von Humboldt's *Kosmos* was in publication, and nearly burned her manuscript. Instead, astronomer John Herschel, Caroline's nephew, convinced her to release her work regardless. Somerville's book was not only well received by the public and scientific community, but Humboldt himself praised her work (Brück 2009).

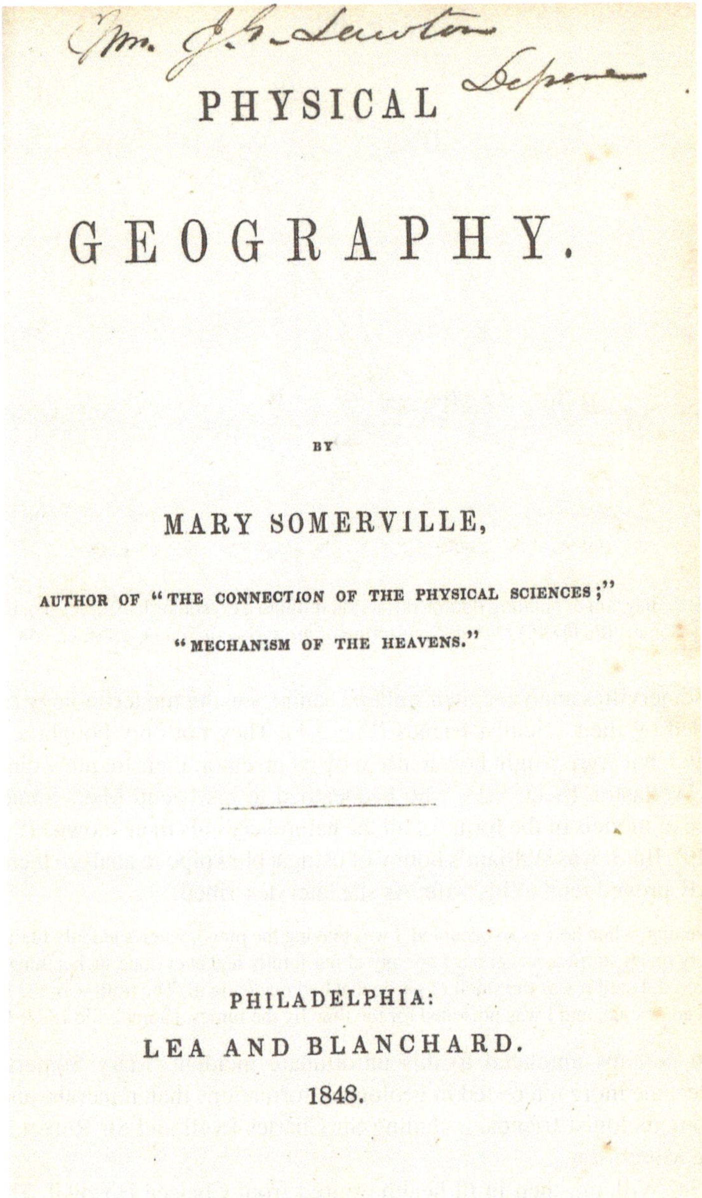
PHYSICAL

GEOGRAPHY.

BY

MARY SOMERVILLE,

AUTHOR OF "THE CONNECTION OF THE PHYSICAL SCIENCES;"

"MECHANISM OF THE HEAVENS."

PHILADELPHIA:

LEA AND BLANCHARD.

1848.

Fig. 9.2 Title page. Reproduced from Mary Somerville (1848) *Physical Geography,* Amer. edn

Physical geography was still a developing field at the time, which allowed Somerville to define it in her own way. In doing this she used a decidedly broad brush, as her work included not only topics in geology and geography, but meteorology, climatology, thermodynamics, and magnetism. In his review of her work, Henry Holland (1848: 193) defined it as "that branch of science which embraces all matter, in all its forms of existence, organized or inorganic, forming the great globe on which we dwell," while also making the error of referring to her work as "the first English work bearing the title, and distinctly comprehending what belongs to this great subject" (Holland 1848: 193). Mary herself only included a concise definition for her readers in later editions, offering that

> Physical Geography is a description of the earth, the sea, and the air, with their inhabitants animal and vegetable, of the distribution of these organized beings, and the causes of that distribution. Political and arbitrary divisions are disregarded, the sea and the land are considered only with respect to those great features that have been stamped upon them by the hand of the Almightly…. (Somerville 1855: 13).

Creese and Creese (1998) consider the work revolutionary in its disregard for political boundaries, and for avoiding the usual method of discussing geography as a country by country survey.[4] However, a similar architecture was developed several decades prior by Emma Willard for her geography curriculum (Sect. 8.3).

But Mary Somerville was not unconventional in all respects. According to her daughter, Martha, she "was profoundly and sincerely religious; hers was not a religion of mere forms and doctrines, but a solemn deep-rooted faith which influenced every thought, and regulated every action of her life" (Somerville 1874: 374). Martha Somerville details how her mother found no conflict between her personal beliefs and the geological record. On the contrary, she found

> The theories of modern science… quite in accordance with her religious opinions. She rejected the notion of occasional interference by the Creator with His work, and believed that from the first and invariably He has acted according to a system of harmonious laws, some of which we are beginning faintly to recognise, others of which will be discovered in course of time, while many must remain a mystery to man while he inhabits this world. (Somerville 1874: 375)

This personal philosophy is clearly reflected in the opening pages of *Physical Geography*, where she explains that floods and earthquakes

> the august and terrible ministers of Almighty power, have torn the solid earth and opened the seals of the most ancient records of creation…. There we read of the changes that have brought the rude mass to its present fair state, and of the myriads of beings that have appeared on this mortal stage, have fulfilled their destinies, and have been swept from existence to make way for new races which, in their turn, have vanished from the scene till the creation of man completed the glorious work. Who shall define the periods of those mornings and evenings when God saw that his work was good?…These stupendous changes may be but cycles in those great laws themselves and He who has ordained them. (Somerville 1848: 2–3)

Yet these rather liberal views were not universally shared by all in mid nineteenth century England, and after the publication of *Physical Geography* Mary Somerville was the victim of a public rebuking during a sermon at York Cathedral (Somerville 1874: 129). It is therefore ironic that in a sketch of Somerville's life, Sarah Josepha Hale proclaimed that in her ability to put forth works of "science *conjoined with moral advancement*…. Mrs. Somerville has done more by her writings to Christianize the Sciences than any living author… Physical science will, henceforth, have a religious power" (1853: 790; emphasis original). However, as

[4] For a close analysis of Somerville's *Physical Geography*, see Neeley (2001).

Somerville understood, regardless of one's personal beliefs, "facts are such stubborn things (1874: 129).

Grace Anne Milne (later Lady Prestwich, wife of geologist Joseph Prestwich), described the experience of accompanying her uncle, paleontologist Hugh Falconer, to visit Somerville in 1859 as being "in the presence of an uncommon woman. There was no mock humility, no false modesty, but there was the stamp of dignity on her features, and, if I might use a long word, a self-containedness that could not fail to impress those who came in contact with her" (Prestwich 1901: 123–4). Somerville's then decidedly uncommon and successful career as a translator of science for an adult audience presaged the eventual decline of the familiar format, and the transition to a more modern style of science popularization.

References

Baily F (1838) On the remarkable phenomenon that occurs in total and annular eclipses of the sun. Memoirs RAS 10:1–40
Brück MT (2009) Women in early British and Irish astronomy. Springer, Dordrecht
Creese MRS, Creese TM (1998) Ladies in the laboratory? Scarecrow Press, Lantham
Creese MRS, Creese TM (2004) Ladies in the laboratory II: Western European women in science, 1800–1900. Scarecrow Press, Lanham
Denham G (1815) Remarks on the writings and prophecies of Joanna Southcott. Dean and Munday, London
Fell-Smith C (1896) Roberts, Mary (1788–1864). In: Lee S (ed) Dictionary of national biography v48. MacMillan and Co., New York
Gage WL (1856) Preface to the American edition. In: Zornlin RM (ed), Physical geography for families and schools, rev edn. James Munroe and Co, Boston, pp xiii–xiv
Gould SJ (1997) The invisible woman. In: Gates BT, Shteir AB (eds) Natural eloquence: women reinscribe science. Wisconsin UP, Madison
Hale SJ (1853) Woman's record; sketches of all distinguished women. Harper, New York
Holland H (1848) Physical geography by Mary Somerville. Littell's Living Age 20:193–209
Leach C (2006) Religion and rationality: Quaker women and science education 1790–1850. Hist Educ 35(1):69–70
Lightman B (2007) Historians, popularizers, and the Victorian scene. In: Lightman B (ed) Victorian popularizers of science: designing nature for new audiences. Chicago UP, Chicago, pp 1–38
Miller H (1841) The old red sandstone; or new walks in an old field. Johnstone, Edinburgh
Mitchell M (1860) Mary Somerville. Atlantic Monthly 5:568–571
Morgan C (1842) What is a voltaic battery? by Rosina M. Zornlin. The Athenaeum no 778. pp 833–834
Neeley K (2001) Mary Somerville: science, illumination, and the female mind. Cambridge UP, Cambridge
O'Connor R (2007) Young-earth creationists in early nineteenth-century Britain? towards a reassessment of 'scriptural geology. Hist Sci 45:357–403
Patterson EC (1969) Mary Somerville. Br J Hist Sci 4(16):311–339
Patterson EC (1983) Mary Somerville and the cultivation of science 1815–1840. Martinus Nijhoff, The Hague
Prestwich GA (1901) Essays descriptive and biographical. Blackwood, Edinburgh
Roberts M (1824) The wonders of the vegetable kingdom displayed in a series of letters, 2nd edn. Whittaker, London

Roberts M (1838) The progress of creation, considered with reference to the present condition of the earth. Smith, Elder and Co., London
Shteir AB (1996) Cultivating women, cultivating science. Johns Hopkins UP, Baltimore
Somerville M (1848) Physical geography, v1. Murray, London
Somerville M (1855) Physical geography, Amer. edn. Blanchard and Lea, Philadelphia
Somerville M (ed) (1874) Personal recollections, from early life to old age, of Mary Somerville. Murray, London
The Dublin Review (1838) Review: religious novels. 5:533–555
The Monthly Review (1836) The solar eclipse; or, the two almanacks. by R.M. Zornlin. 2(1):185–186
The North American Review (1856) Physical geography for families and schools by R.M. Zornlin. 82:281
The Spectator (1835) Children's books. 8:760
Thomas R (1905) Zornlin family. Notes Queries 10(3):402–404
Urdank AM (1990) Religion and society in a Cotswold vale: Nailsworth, Gloucestershire, 1780–1865. California UP, Berkeley
Zornlin RM (1839a) Recreations in geology. Parker, London
Zornlin RM (1839b) Observations of shooting stars on the nights of the 10th and 11th of August, 1839. Phil Mag 15:441–442
Zornlin RM (1840) Preliminary discourse on the nature and advantages of geology. Satur Mag 16:11–12
Zornlin RM (1841) On the periodical shooting stars, and on shooting stars in general. Phil Mag 19:347–357
Zornlin RM (1842) Recreations in physical geography, or, the earth as it is, 2nd edn. Parker, London
Zornlin RM (1843) What is a voltaic battery? Parker, London
Zornlin RM (1852) Outlines of geology for families and schools. Parker, London
Zornlin RM (1859) On heat and on the indestructibility of elementary bodies. Report of the 28th meeting of the British Association for the Advancement of Science Part II: notes and abstracts of miscellaneous communications to the sections: mathematics and physics. Murray, London, p 6

Chapter 10
The Extinction of the Familiar Format

10.1 Overview

In 1866 noted art and social critic John Ruskin (1819–1900) published a dialogue in mineralogy entitled *The Ethics of the Dust. Ten lectures to Little Housewives on the Elements of Crystallisation.* In his preface he claims

> The following lectures were really given, in substance, at a girls' school (far in the country); which, in the course of various experiments on the possibility of introducing some better practice of drawing into the modern scheme of female education, I visited frequently enough to enable the children to regard me as a friend. … It will be at once seen that these Lectures were not intended for an introduction to mineralogy. Their purpose was merely to awaken in the minds of young girls, who were ready to work earnestly and systematically, a vital interest in the subject of their study. (Ruskin1908: 3–4)

In a script like style written in stilted language, Ruskin begins to describe a "very idle talk, by the dining-room fire, after raisin-and-almond time" between an "Old Lecturer" and several young girls. The lecturer asks them to describe "what the make-believe was, this afternoon," to which Isabel (after "arranging herself very primly on the foot stool") answers that it was about being "lost in the Valley of Diamonds" (1908: 11). The lecturer transforms this into a discussion about the source of diamonds in rocks, and he allows the girls to hold a rock that has gold and diamonds embedded in it while he admonishes them about the dangers of these

> fatal jewels; native here in their dust with gold, so that you may see, cradled here together, the two great enemies of mankind, - the strongest of all malignant physical powers that have tormented our race…. Was any woman, do you suppose, ever the better for possessing diamonds? but [sic] how many have been made base, frivolous, and miserable by desiring them? Was ever man the better for having coffers full of gold?.... The sin of the whole world is essentially the sin of Judas. Men do not believe their Christ; but they sell him. (Ruskin 1908: 18–9)

But while diamonds are evil, there are crystals that possess positive virtues that can be seen in their "fighting their faults. And some have a great many faults; and some are very naughty crystals indeed…. crystals have a limited, though a stern, code of

K. Larsen, *The Women Who Popularized Geology in the 19th Century*,
DOI 10.1007/978-3-319-64952-8_10

morals; and their essential virtues are but two; - the first is to be pure, and the second is to be well shaped" (Ruskin 1908: 69–70). The metaphor is clear, in terms of the moral lesson being taught to the impressionable young girls. The titular "dust" is just another name for atoms, or more generally base matter (as well as a reference to the famous admonition from Genesis concerning the impermanence of the flesh). It also signifies the traditional female role of housekeeper through which Ruskin is attempting to interest his young charges in what he suspects might otherwise be an uninteresting topic, the crystallisation of minerals (Myers 1989).[1]

The critical response to Ruskin's novel was swift and unequivocal, as noted in his new preface to the second edition (published in 1877): "I have seldom been more disappointed by the result of my best pains given to any books, than by the earnest request of my late publishers, after the opinion of the public had been taken on the Ethics of the Dust, that I would 'write no more in dialogue!' However, I bowed to public judgment in this matter at once" (Cook and Wedderburn 1905: 203). In explaining his decision to violate this request in publishing a second edition, Ruskin blames a friend who had requested it to be reissued, and asks "the readers whom it may at first offend by its disconnected method, to examine, nevertheless, with care, the passages in which the principal speaker sums the conclusions of any dialogue: for these summaries were written as introductions, for young people" (Cook and Wedderburn 1905).

Nearly 50 years before Ruskin's ill-fated publication of *The Ethics of the Dust*, Robert Bakewell had noted in the preface to *An Introduction to Mineralogy* that "He who consults his own reputation more than the benefit of his reader, will scarcely choose to write on subjects of science in the way of dialogue, as this form has fallen into dispute, since the admirable dialogues of Berkley" (1819: vii). Indeed, while Marcet's famous *Conversations on Chemistry* had sparked an interest in the familiar format, by the 1830s the genre was in noticeable decline, and by mid century it had disappeared from botany, the field in which it had made such a strong impact in the late eighteenth century (Shteir 1996). The precipitous decline in the popularity of the familiar format was caused by a number of factors related more to society than science itself. Due to the inclusion of a familial setting and strong maternal characters, and, importantly, the gender of most of the authors themselves, the familiar format became synonymous with the feminine, the amateur, and the old-fashioned (Shteir 1997). As was seen in the case of Almira Phelps (Sect. 8.4), this was already becoming problematic circa 1830.

In some ways, the very success of the women writers themselves became part of the problem, as both male writers and men of science began pushing back against the influx of women into what was deemed their territory. The first science in which such a concerted effort was waged was, predictably, botany. Shteir (1996) describes how botany became more masculinized simultaneously with becoming more professionalized. But a number of these writers were not satisfied to merely take control of the dialogue of botany (ironically, by shutting down the dialogue format of writing within the field), but concurrently ridiculed both the format and the women

[1] For analysis of *The Ethics of the Dust*, see Myers (1989).

writers they sought to replace. A commonly cited example is John S. Forsyth's *The First Lines of Botany* (1827), written not in the form of a dialogue, but rather as a "more intelligible narrative" (17). Not only does the work's introduction contain (as noted in Sect. 5.3) a scathing condemnation of the "baby" or "Interrogative" style (15), but Forsyth condemns as even more loathsome works written in the

> conversational or epistolary style, by people indifferently acquainted with the sciences they would thereby promulgate; and even in the best hands, there is something in this method at which the unassuming, industrious, and inquisitive mind recoils; and more particularly when forced, as it were, to draw, through such equivocal channels, upon the fictitious correspondence of some garrulous old woman or pedantic spinster, for the higher order of elementary knowledge. (17)

The stereotypical identification of writers of this style with old, unmarried women is curious, as the most famous author of this genre, Mrs. Jane Marcet, was both married and hardly old when she published her first work, and Delvalle Lowry was a young woman.

As described in Chap. 8, it became more common for professors to write textbooks for their own and others' classes, and since the overwhelming majority of professors were male, it was inevitable that they would define what would become the standard format. This style was strictly utilitarian, focused on the dissemination of scientific knowledge rather than ethics and religious faith. The emphasis on actively engaging the reader disappeared, and with it the need for the ancillary scaffolding structure of the fictional characters and their familial setting (Shteir 1996). In addition to competition from professors and practicing scientists, women writers also faced competition in the form of public education. Homeschooled children and their mothers had been the original intended audience of this genre of works. With the increasing availability of both public education and science courses within such schools the familiar format became doomed for extinction (Gates 1998). Finally, while the familiar format's accidental audience, adults with no formal education in science, had not decreased, publishers had become savvy to both their interest and economic value. In the second half of the nineteenth century popular level magazines such as *Knowledge* (1881), *Popular Science* (1872), and *Popular Astronomy* (1893) fed the public's interest in science. Newspapers often carried articles on scientific discoveries as well. But most notably, scientists themselves took notice of the public's interest in their fields, and as Peter Bowler (2009) documents, by the early twentieth century it was common for illustrious British scientists of the day (including J.B.S. Haldane and Sir Arthur Eddington) to write popularized works.

But women continued to write popularized works of geology and other sciences in the late nineteenth century, especially, but not exclusively, for children, despite both the competition from professional scientists, and the public's expectation for new literary styles in such works. In the case of some of these women, their creativity led them to new heights of commercial success within the now male-dominated world of popular level science. The stories of three of these women will be summarized, painting them and their works as a bridge between the old century and the new.

10.2 Arabella Buckley (1840–1929): Entering *The Fairy-land of Science*

Charles Lyell had played a pivotal role in the transition from a Biblical natural history to a modern geology. It is perhaps ironic that his secretary, Arabella Buckley (1840–1929), likewise navigated a liminal position in the history of the popularization of science. Arabella was born on October 24, 1840 in Brighton, England, the youngest daughter of the Reverend John Wall Buckley and his wife Elizabeth Burton (d. 1889). Her father served for 40 years as the Vicar of St. Mary's, Paddington. Arabella had at least four brothers, including Henry Burton Buckley (1845–1935), a barrister and judge who became the first Baron Wrenbury in 1915. Buckley served as Charles Lyell's secretary from 1864 until his death in 1875, and by her own description, through this "privilege… was thus brought in contact with many of the leading scientific men of our day" (Buckley 1876: v). One of these luminaries was Charles Darwin, whom she met c.1863, and who encouraged her to write children's books on animals (Osborn 1926). She took up writing shortly after Lyell's death, and her first book, *A Short History of Natural Science* (1876), was dedicated to the memory of "my beloved and revered friends Sir Charles and Lady Lyell to whom I own more than I can ever express... trusting that it may help to develope [sic] in those who read it that earnest and truth-seeking spirit in the study of God's works and laws which was the guiding principle of their lives" (Buckley 1876: i).

In her preface Buckley explains that her intention as an author is "to place before young and unscientific people those main discoveries of science which ought to be known by every educated person, and at the same time to impart a living interest to the whole, by associating with each step in advance some history of the men who made it" (1876: v). She invokes her personal experience as Lyell's secretary in further explaining that "I often felt very forcibly how many important facts and generalizations of science, which are of great value both in the formation of character and in giving a true estimate of life and its conditions, are totally unknown to the majority of otherwise well-educated persons" (1876: v–vi). In this invocation of the role of science in the teaching of ethics and morality, Buckley is clearly echoing her foremothers in the familiar format, demonstrating that, while the style of writing might have shifted over the decades, some of those earlier core values still remained. But just how much had women's science writing changed from the time of the familiar format's heyday?

In the case of Buckley, the answer was that it had, in fact, not shifted in intent, just mechanics, and even then, the bones of the familiar format were still plainly visible. Her work lays out important scientific and mathematical discoveries, beginning with the Ancient Greeks. She speaks to the reader in a conversational way, comparing the study of stars and plants to learning about a group of children attending a school. It is not merely important to learn their names, but also "their character, their habits, and other real facts about them" (Buckley 1876: 2). She similarly explains to her reader that the goal of her work is to illuminate "how science has *become* what it is" (1876: 3). Her description of Lyell and his place in the history of science is, as expected, full of praise and perhaps a touch of hyperbole. She says, in part:

> Sir Charles Lyell taught men to read the true history of the earth. It is difficult in the present day to understand rightly how great a work he accomplished, for though his ideas were ridiculed in the beginning, yet he lived long enough to see all men agree with him, and his doctrines received as self-evident truths. Like all other great men, he was humble and reverent in his study of nature. His one great desire was to arrive at truth, and by his conscientious and dispassionate writings he did much to persuade people to study geology calmly and wisely, instead of mixing it up with angry disputes, like those which, in the time of Galileo, disfigured astronomy. (1876: 409)

In concert with the book she gave lectures on natural science from 1876–83 and continued to write science books for children, including *The Fairy-land of Science* (1878), *Life and her Children* (1881), *Winners in Life's Race; or, the Great Backboned Family* (1883), *Through Magic Glasses* (1890), *Moral Teachings of Science* (1891), and *Eyes and No Eyes* (1901). She was also the editor of an 1877 edition of Mary Somerville's *Connexion of the Physical Sciences* and the picture book *Animals from Life* (1887) based on illustrations by German illustrator Heinrich Leutemann, for which she also translated the text.

The anonymous reviewer of her *The Winners in Life's Race* in *The American Naturalist* calls the work "the most successful attempt at a popular sketch of modern zoology with which we are acquainted," noting that she "wields an easy and graceful pen" (1884: 47). Her later works are of particular note because they were published under her maiden name even after her 1884 marriage to New Zealand native physician and widower Thomas Fisher (c. 1819–1895). For example, a print ad for *Animals from Life* in *The Atheneum* touts the work as "A picture book, edited by Miss Buckley" and cites positive reviewers who refer to her alternately as "Miss Buckley" and "Mrs. Fisher" (1887: 690).

In her biographical entry in *Who's Who 1907* she lists her recreations as simply "reading, patience" (Black and Black 1907: 599). However, she also had a longstanding interest in spiritualism that she kept hidden from most of her scientist friends for fear of it coloring her reputation as a science writer. A notable exception was evolutionist Alfred Russel Wallace, whose shared interest in the topic became the basis of a deep friendship (Lightman 2009).[2] As part of her investigations into spiritualism, Buckley visited mediums and believed that she had had contact with a dead sister. Her spiritualist beliefs, as well as her preference to consider evolution in terms of mutualism rather than competition (a reflection of her religious faith in a benevolent Divine Power whose design was evident in the workings of nature), deeply impacted her personal views on evolution, and are reflected in her books. She died at her home in Devon on February 9, 1929, the cause of death being influenza.

[2] For more information on Buckley's spiritualism and her relationship with Wallace, see Lightman (2009: 238–53).

This complex system of beliefs is reflected in her most famous work, *The Fairy-land of Science*, which was published both in England and America through 1905. Her goal is to paint science as a magical fairy-land, more wondrous and fascinating than any imaginary world found in myths, fables, or legends, and thereby interest her readers in the natural rather than fictional world.[3] Buckley is aware that her reader will be skeptical at first, calling it a

> somewhat bold promise, seeing that most of you probably look upon science as a bundle of dry facts, while fairy-land is all that is beautiful, and full of poetry and imagination. But I thoroughly believe myself, and hope to prove to you, that science is full of beautiful pictures, of real poetry, and of wonder-working fairies; and what is more, I promise you they shall be true fairies, whom you will love just as much when you are old and greyheaded as when you are young.... and though they themselves will always remain invisible, yet you will see their wonderful power at work everywhere around you. (1880: 1–2)

The power referenced here is, of course, the Deity's.

Buckley begins to weave her literary spell, explaining that her fairies are the forces of nature, and just as the fairies of fiction are invisible (except to those who are trained to see them), so, too, are the forces of the world. The way to enter the fairy-land of science and see the forces in action is through the power of imagination, specifically the ability of children to make "pictures or *images* in our mind, of that which *is*, though it is invisible to us" (1880: 7). Buckley begins to introduce the reader to her scientific fairies, piquing their interest with a mere taste of their individual powers, including "fairy *Cohesion* ever ready to lock atoms together when they draw very near to each other: or fairy *Gravitation* dragging rain-drops down to the earth: or the fairy *Crystallization* building up the snow-flakes in the clouds.... Do you care to know how another strange fairy, *Electricity*, flings the lightning across the sky and causes the rumbling thunder?" (1880: 12). Having hooked her audience, she now prepares to reel them in by asking "But if you ask yourself why things happen, and how the great God above us has made and governs this world of ours; if you listen to the wind, and care to learn why it blows.... you will learn to know and love those fairies" (1880: 13). She returns to this deeply spiritual reason for studying science more than once in her book, for example offering rather poetically

> We are all groping dimly for the Unseen Power, but no one who loves nature and studies it can ever feel alone or unloved in the world. Facts, as mere facts, are dry and barren, but nature is full of life and love, and her calm unswerving rule is tending to some great though hidden purpose. You may call this Unseen Power what you will – may lean on it in loving, trusting faith, or bend in reverent and silent awe; but even the little child who lives with nature and gazes on her with open eye, must rise in some sense or other through nature to nature's God. (1880: 25)

Buckley's understanding of the supernatural world does not conform with that of orthodox Christianity, but is colored by her spiritualist interests. One could even argue that it has pantheistic tendencies. Just as the classification of minerals had proven problematic a century before, forcing Buckley's personal experience of the divine into a standardized taxonomy is fraught with frustrations and will not be

---

[3] For a close analysis of *The Fairy-land of Science* and Buckley's other works see Gates (1998).

attempted here. It is sufficient to acknowledge that in her belief in a deeply spiritual connection between the divine and the natural world, Buckley harkens back to an earlier type of writing.

In terms of her literary style, as in the case of *A Short History of Natural Science*, Buckley is employing a one-sided conversational tone (for example, addressing the reader as *you*) that is reminiscent of the familiar format. She is also utilizing everyday examples (common phenomenon such as weather and gravity) and referencing the moral/religious benefits of studying science. One brief example is her lecture on "The Two Great Sculptors – water and ice," where she compares the action of water and ice upon the surface of the earth to the way that "the wrinkles and curves of a statue are cut by the scuptor's chisel," resulting in the creation of "the hills and valleys, the steep slopes and gentle curves on the face of our earth" (1880: 100). A more detailed lesson is found in the "History of a piece of coal," which starts with "those plants and trees of the far-off ages, which seemed to lead such useless lives" and now fuel the fires of industry (1880: 192). She uses this metamorphosis as a moral as well as scientific lesson, noting

> There are many people in the world who complain that life is dull, that they do not see the use of it, and that there seems no work specially for them to do. I would advise such people, whether they are grown up or little children, to read the story of the plants which form the coal. These saw no results during their own short existences, they only lived and enjoyed the bright sunshine, and did their work, and were content. And now thousands, probably millions, of years after they lived and died, England owes her greatness, and we much of our happiness and comfort, to the sunbeams which those plants wove into their lives. (Buckley 1880: 192)

There is also a greater spiritual lesson that takes on a deep meaning in the context of Buckley's belief in communication with the dead: "They burst forth again in our fires, in our brilliant lights, and in our engines, and do the greater part of our work, teaching us 'That nothing walks with aimless feet/ That not one life shall be destroyed,/ Or cast as rubbish to the void,/ When God hath made the pile complete'" (Buckley 1880: 192).

Finally, she also engages in another key trope associated with the familiar format, negotiating with her audience exactly which adaptations she is willing to make for their benefit, and which she is not. Buckley is clear in her expectation that her reader

> must learn something of the language of science. If you travel in a country with no knowledge of its language, you can learn very little about it: and in the same way if you are to go to books to find answers to your questions, you must know something of the language they speak. You need not learn hard scientific names, for the best books have the fewest of these, but you must really understand what is meant by ordinary words. (Buckley 1880: 14–5)

As examples she discusses the difference between solids, liquids, and gases. Buckley is cognizant of the possibility of introducing scientific errors through oversimplification and therefore stresses both scientific accuracy as well as what she sees as the inherent wonders of the natural world, unwilling to sacrifice one for the other. On the contrary, she argues that this "is the reason why so many complain that science is dry and uninteresting. They forget that though it is necessary to learn accurately, for so only we can arrive at truth, it is equally necessary to love knowledge and make it lovely to those who learn, and to do this we must get at the spirit which lies under the facts" (Buckley 1880: 19). Her main point here is the same as that stressed in *A Short History of Natural Science* in her example of the stars and the school children: "It is useful to be able to classify a flower and to know that the buttercup belongs to the Family Ranunculaceae…. But it is far sweeter to learn about the life of the little plant, to understand why its peculiar flower is useful to it, and how it feeds itself, and makes its seed. No one can love dry facts; we must clothe them with real meaning and love the truths they tell, if we wish to enjoy science" (Buckley 1880: 19–20).

In the end it is clear that in the case of Buckley's writings, the familiar format has been renegotiated, but much of its central tenets remain. Gone are the active hands-on experiments, dissections, and walks devoted to mineral and fossil collecting, but they are replaced by a much greater emphasis on observation and an understanding of the nature of science as a process. It has moved beyond the confines of the domestic sphere, into the larger world of experience, and while the maternal guide has become perhaps more distant, she remains in the voice of the female author/narrator who actively guides her reader to an understanding of the natural world, and through it, the divine.

10.3 Agnes Giberne (1845–1939): The Lingering Influence of Religion on Geology

Buckley had not taken up her pen in the name of science popularization until in her 30s, and then found success that lasted for several decades. Agnes Giberne (1845–1939) likewise did not publish her first science popularization until age 35, but had an active career as a writer of various genres that both began far earlier and lasted until her death.

Agnes, who was descended from an French aristocratic family, the de Gibernes, was born in Ahmednugger, India, to Major Charles Giberne, of the Indian Army, and his wife Lydia Mary Wilson. As a young child she moved to England upon her father's retirement from the military, and was educated at home. She attributes her literary tastes to her mother and her scientific proclivities to her father (Black and Black 1907). She began to "scribble stories" at age seven, which she originally only shared with her sisters, and published her first children's story, anonymously, 10 years later through the S.P.C.K. (Society for Promoting Christian Knowledge), an Anglican mission organization (Black and Black 1907: 672). Known for their religious spirit, her stories appeared in both book form and serialized in magazines, especially those with a religious audience, such as *Sunday at Home* (published by the Religious Tract Society) and the *S.P.C.K Magazine*. In the 1860s and 1870s she was regularly publishing stories at the rate of several per year (*The Times* 1939).

She began publishing popularizations of science in 1880, mostly in astronomy, but also to a lesser extent in meteorology, oceanography, botany, and geology. Her astronomical works[4] included *Sun, Moon, and Stars: Astronomy for Beginners* (1880) and its sequel for younger children *Among the Stars, or Wonderful Things in the Sky* (1885), *Radiant Suns* (1894), *The Starry Skies* (1894), *The Story of the Sun, Moon and Stars* (1908), and *This Wonderful Universe* (1923). All of her science works contain the same religious themes as her strictly fictional writing, with a strong sense of evangelism (Daw 2014). Her deep interest in astronomy is also reflected in her work in amateur astronomy organizations. When the gender-inclusive Liverpool Astronomical Society folded in 1890 due to financial and administrative difficulties, Edward Walter Maunder of the Royal Observatory (Greenwich), and others founded a new amateur society, the British Astronomical Association (BAA). A Provisional Committee was set up in 1890, which included four women: solar observer Elizabeth Brown, science writers Agnes Clerke and Agnes Giberne, and Margaret Huggins (1848–1915), who had conducted seminal research on astronomical spectroscopy and photography with her husband, William. Women filled positions of authority from the beginning, with Brown becoming the Director of the Solar Section and Clerke and Huggins elected to Council seats.[5] In her biographical note in the 1907 edition of *Who's Who* Agnes' hobbies are listed as "reading, music, [and] open-air exercise," which may have helped her live to the age of 94 (Black and Black 1907: 672).

In her preface to *Sun, Moon, and Stars*, Giberne explains that she means for the work to serve

> not merely a text-book for school use, though possibly it may serve for that also, but a volume of sufficient interest for general reading; not merely a book adapted for intelligent boys and girls, though I hope it may be found serviceable for them also, but a volume fitted for "beginners" of all kinds, whether children, working-men, or even grown people of the educated classes, who should have a desire to enter for the first time on the study of this fascinating science. (1879: xi)

Note that she is not limiting her audience to children, but rather specifically targeting beginners of all ages. But there is a second admitted reason for issuing this work, namely, as in the case of works of half a century prior, to be of religious service. As she explains "In the Book of Nature, side by side with the Book of Revelation, we may learn some things about our Father in Heaven. It would be happiness to feel that I had helped any, however slightly, to look upward through Nature unto Nature's God" (1879: xii). But whereas Buckley's work appears to represent a modification of the familiar format, Giberne's works, can perhaps best be thought of as occupying a transitional space between the familiar format and the more formal lecture/textbook style that was represented in the works of Mary Somerville and was increasingly becoming the norm with the masculinization and professionalism of the popularization of science. For example, *Sun, Moon, and Stars* is written in a conversational tone utilizing the pronouns *I* and *We*, but it is more formal and

[4] For a deeper analysis of her astronomical works, see Daw (2014).

[5] For more information on the role of women in the BAA see Larsen (2006).

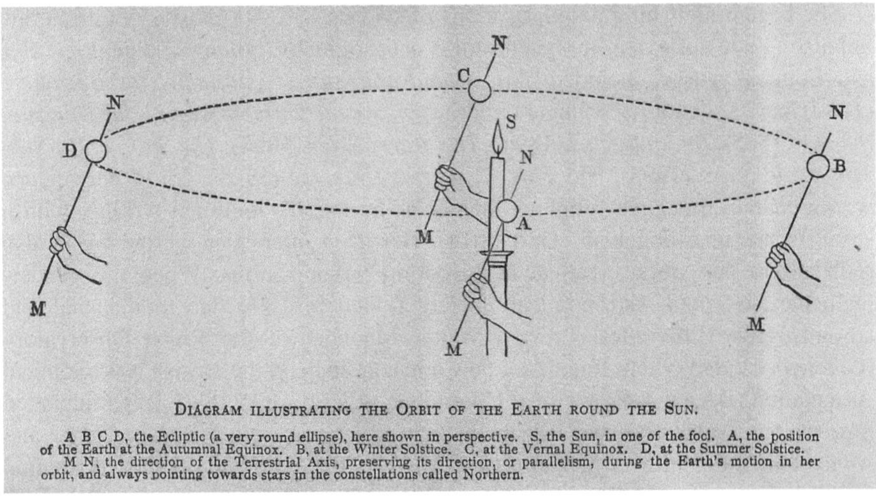

Fig. 10.1 Diagram of an experiment using an orange on a knitting needle and a candle to represent the tilted earth orbiting the sun. Reproduced from Agnes Giberne (1879) *Sun, Moon, and Stars*

distant; one does not get the sense that she is relating to the audience on a personal level, as a female tutor or teacher, as in the case of Buckley's work. There are hands-on demonstrations suggested, for example when she directs the reader to stick a knitting needle through an orange to represent the tilted earth (Fig. 10.1) and have it orbit around a candle, representing the sun, but overall the descriptions are more reminiscent of a textbook than a storybook.

The sequel, *Among the Stars or Wonderful Things in the Sky* (1885) is aimed at a younger audience and, correspondingly, is written in a different style. She suggests in her preface that the "want has been repeatedly suggested to me of a little volume for children on the subject of Astronomy, —'*much easier than* "Sun, Moon, and Stars"'" (Giberne 1885: vii; emphasis original). In recognition of the "warm reception given to that former work of mine, I now venture to offer such a little book to friends and readers" to fulfill "a household need" (Giberne 1885: vii). This specific invocation of the domestic sphere is just one line of evidence that ties this work to a style that, like Buckley's *The Fairy-land of Science*, incorporates many, but not all, of the defining characteristics of the familiar format, despite its appearance so late in the nineteenth century. By that time this would have been considered an antiquated style, and was presumably only accepted without ridicule because of the young readership for which it was intended. Though a series of conversations, letters, and a fantastical story within a story created for the main character (in which he visits the moon on the "Wings of Imagination" [1885: 152]), a curious young boy named Ikon learns about astronomy (Fig. 10.2). Although the story is replete with religious overtones, written in the literary style of conversations and letters, and takes place in a familial setting, what is decidedly missing from this work is the female expert who imparts knowledge. Indeed, the only major female character in the work is Ikon's nurse, Dormer, who is incapable (as is Ikon's father) of answering

Fig. 10.2 Frontispiece showing the boy Ikon observing a star. Reproduced from Agnes Giberne (1885) *Among the Stars*

the child's questions. Instead, the expert/teacher role is filled by Professor Herr Lehrer, a friend of the boy's father, and Lehrer's friend Mr. Fritz, the owner of a large telescope that Ikon is able to look through (as well as the creative author of the fictional moon adventure).

Between these two astronomical works Giberne turned her pen to a different science but the same audience as the first work. The preface to *The World's Foundations or Geology for Beginners* (1882) explains that the "very warm and hearty reception accorded to my little book on Astronomy, has been my best encouragement in entering upon the domain of the sister science, Geology" (1882: v). What she refers to as a "companion-volume" to *Sun, Moon and Stars* is "written upon much the same plan, and is intended for the same class of readers – for Beginners of all kinds, whether poor or rich, whether boys, girls, or grown-up people" (Giberne 1882: v). She claims that her main reason for producing this work is to "open the eyes of others to the hidden wonders and possibilities of enjoyment which lie folded in this little-studied branch of science" (Giberne 1882: v). This is a curious claim, indeed, given the large library of popular-level works on the science that had been produced in preceding decades (and described within this book) and the fact that geology had become a standard subject in schools alongside astronomy. But perhaps Giberne was referring to a paucity of people engaging in free-choice informal learning in their leisure time, because "Geology is counted by many to be a dull subject. But if it has its dry bones, it has also its forms of poetic beauty, its scenes of loveliness, its chords of sublime harmony" (Giberne 1882: v). Note the return to a romantic, poetic view of geology reminiscent of numerous previous authors.

But the devoutly religious Giberne also clearly warns that "Geology is counted by others to be a dangerous subject," referring to the potential for geology to disagree with Scripture. But Giberne offers that "if so, the danger lies in ourselves, not in Geology. Man's haste in decision, and his readiness to put faith in unproved theories, may lead him astray. The study of God's truths, if rightly undertaken, cannot cause his feet to wander" (1882: v–vi). As was commonly asserted by most of the earlier authors examined in this work, Giberne argues that "Geology speaks to us, as surely as the Bible itself speaks to us, of the Creator and His ways, albeit in terms more ambiguous, in language more easily misunderstood. The one is His Word, the other is His Handiwork. That the one should contradict the other is not possible" (1882: vi), a claim that has been noted previously in other works. For example, compare this to the assertion by Delia Godding that "nature's works and God's word agree in their testimony" (1847: 128).

While Giberne does not list her references in the preface, she claims that the "leading Geological writers of England and of America have been my authorities," again a revisiting of a common method of establishing authorial authority utilized by women earlier in the century (1882: vi).

The work is divided into three parts: geological materials (including fossils), a reconstruction of the geological past, and ongoing geological processes (including volcanoes and the work of rivers). Giberne explains that while her book "is not a strictly religious work… no volume on the subject of Geology can be fairly and honestly written without frequent reference to the Divine Architect of the great Earth-crust Building" (1882: 209). She is unwavering in this conviction; each chapter begins with a relevant quote from the Bible, and religious metaphors, such as rocks and fossils as "marvellous [sic] tokens of the goodness and greatness of God," abound (1882: 3) Each of the three sections of the book ends with a chapter or chapter section tying the geological evidence back to religious truth, reminiscent of the religious reflections found in Jane Kilby Welsh's *Familiar Lessons in Mineralogy and Geology* (Sect. 5.2). For example, the section on the materials of the earth ends with a chapter entitled "The Two Books," a reference to the Bible and the "many-volumed Book of Nature, to which belongs the torn and battered volume of Geology" (Giberne 1882: 87). Giberne explains to her reader that the interpretation of the six days of creation as "six *short* days… is a human addition" and that although God could have created the various parts of the world instantaneously, "this does not appear to have been His mode of action" (1882: 86–7; emphasis original). But she is steadfast in her belief (revisited from her preface) that "these two Books, both written by the same great Author, cannot in reality contradict one another" (Giberne 1882: 87).

Giberne then sets up the second section of her book on the history of the earth by naming it "the story of a Father's loving foresight, a Father's careful preparation for His children's needs" (1882: 89). She revisits what she accepts as the ultimate agreement between the geological record and the Biblical record at the end of this section, in a chapter called "The Two Records." She admits that although to "reconcile each separate detail as given in the Book of Divine Revelation and as given in the Book of Nature, is a matter not yet possible with our present imperfect knowledge," she is steadfast in her belief that such an explanation should be found, unless the answers lie beyond the ability of humans to comprehend. If that is to be the case, the "grand Truth of either record remains still unshaken" (1882: 209–10). But she goes further than merely suggesting that any improved geological understanding will ultimately be consistent with Scripture, and claims that "the one Record may and should be used as a help to the understanding of the other" (Giberne 1882: 210). Specifically, if the geological record unequivocally determines that "the commonly-received interpretation of certain Bible-words is wrong, then that interpretation may be given up, but the truth of the written Word stands untouched…. We do not know how far his language is literal and how far it is figurative. Both styles are largely used in the Bible" (1882: 209–10). As an example, she returns to her prior discussion of the six days of creation, offering that "We do not know whether, as some think, the Days may mean vast periods or ages of time, or whether, as others think, they may rather refer to an *appearance* of days and nights, caused by a succession of visions sent to Moses, each in turn dawning, brightening, and vanishing in darkness" (1882: 212). This discussion is curious, as it could have been plucked, wholesale, from works written half a century before, and speaks to the fact that the century-long struggle to find a reconciliation between Scripture and science was not considered so trivial as to no longer be an issue for someone encountering geology for the first time, especially a devoutly religious person.

Finally, at the completion of her discussion of processes found in the modern world, she devotes several pages to her culminating message, that "There is and must be very much that we cannot understand in the science of Geology…. We are but spelling out the broken sentences of the rock-volume, written, as it is, in a strange language, with many missing paragraphs. What marvel if we make mistakes" (1882: 321). But Giberne encourages geologists to continue pushing onward in their understanding for the important reason that in "the great Book of nature much may be learnt about the God of Nature… we shall find ourselves better acquainted than before with the boundless power of Him" (1882: 321–2). Thus it appears that Giberne's book has much in common with similar works written decades before, not in its literary mode of delivery, but in its intended effect on the audience in terms of aiding in their understanding of the negotiated relationship between science and religion. An advertisement included inside the volume reprinted from the *Churchman* clearly echoes this assessment, noting that Giberne

> hopes to teach the young, and all beginners in scientific study, whether young or grown-up, that there is a Christian way of looking at the wonders which have been discovered in the heavens and in the earth. In these pages she tells the story which geologists are wont to tell, though never in so simple and clear a manner, and not always in so reverent a spirit. This author has, we are glad to see, made scientific accuracy an important point, so that the book is both wholesome and reliable. (Giberne 1882: np)

This consideration of the religious and the scientific outcomes—in this order—is also a reminder of the prefaces of many of the works considered in earlier chapters, suggesting that there was still a long road to travel before theology and geology could truly be separated in the popular experience.

10.4 Lady Grace Anne Prestwich (1832–99): Echoes of Mary Somerville

While the religious beliefs of individual geologists had played important roles in their interpretation of geological evidence, this was not the sole type of prejudice that could affect the progress of the science. As Lady Grace Anne Milne Prestwich (1832–99) warned in her articles in popular magazines, there was also the problem of scientists sometimes clinging to their preferred hypotheses long after the tide of evidence had turned toward an alternative explanation.

Grace Anne Milne was born in the Scottish village of Kinloss on December 18, 1832, the eldest of the four daughters and one son born to James Milne (1798–1853) and his wife Louisa Falconer (1802–1878). The famous botanist and paleontologist Dr. Hugh Falconer was one of her maternal uncles. She was described by her biographer and youngest sister Louisa Elizabeth (1844–1920) as "a very attractive, bright child, endowed with the sterling qualities of head and heart which made her father esteemed and beloved, and with a share of her mother's beauty and generous qualities" (Milne 1901: 2). She was sent to boarding school at age six, and it was reported that by age 12 she could write out from memory any sermon that she heard (Milne 1901). She also excelled at drawing portraits, a skill that served her well when she drew geological illustrations for first her uncle and later her second husband, geologist Joseph Prestwich (1812–96).

Hugh Falconer had taken a special interest in his intelligent niece, and during his frequent travels to exotic locations took the time to write her letters. In these he discussed not only geology but its relationship to religion. For example, in a January 19, 1854 letter written in Calcutta he offered that "The Almighty has given us reason, and left us, by the adequate exercise of that power, to investigate the laws and order of creation," but bemoaned the current state of geology, comparing it to Galileo's trials two centuries before:

> When the ignorant and bigoted fail in reason and argument, they raise the yell of intolerance and charge the doctrine with *infidelity*..... As regards the creation of the world, the evidence is as clear that millions and millions of years must have elapsed between the first appearance of life on the earth and the present day, as that you and I possess eyes and ears and have a living existence. (Milne 1901: 6)

Falconer's views concerning the accommodation of scripture and geology give us a fascinating glimpse from the other side of the debate, someone who was apparently not concerned with aligning the two at all.

Grace married wine merchant George McCall (1816–56) on October 18, 1854 and bore him a son, James, on November 23, 1855. Six months later she was a childless widow; George passed away in March 1856 from pneumonia, and James died 2 months later from hydrocephalus (Mather and Campbell 2007). Soon after she attended to another maternal uncle, Calcutta merchant Alexander (1797–1857), as he lingered through a long illness and eventually passed away. In response to these tragedies, Grace lamented to her mother-in-law that she felt "so *crushed*" in her grief (Milne 1901: 8; emphasis original). In September 1858 her life changed for the better when Hugh Falconer invited her to accompany him on a prolonged tour of Italy and France, but playfully admonished her to bring "Only two gowns: when the old wears out buy a new one. No band-boxes !" (Milne 1901: 12). She travelled with her uncle on this trip and many others over the next 7 years, acting as his assistant and secretary, and also making sketches of geological sections and fossils. On one trip they passed through Florence and spent time with Falconer's old friend Mary Somerville. When the pair was not traveling, they lived together at Hugh's home in England, where she acted as hostess. The result of this opportunity was not merely an increased firsthand understanding of geology and experience in field work for Grace, but the chance to meet many of the famous geologists of the day, including Charles Lyell. After Hugh died of rheumatic fever on January 31, 1865, Grace remained at his London address with Hugh's brother, Charles Falconer (1804–79), until she married Joseph Prestwich, wine merchant, amateur geologist, and then President of the Geological Society of London, 20 years her elder, on February 6, 1870. Her husband afterwards turned to science full-time, and received an appointment as a professor of geology at Oxford.

During her second marriage Grace began to write both fiction and nonfiction. She penned two novels that were published anonymously, *The Harbour Bar: A Tale of Scottish Life* (1874) and *Enga* (1880), and a series of geology-based articles for two magazines catering to the middle classes, *Every Girl's Magazine* and the more religious *Good Words*. Among these was a series of six chapters on geology serialized in *Every Girl's Magazine* (1880–1). Mather and Campbell (2009) summarize these as including an account of stratigraphy and chapters on the Paleozoic, Mesozoic, and Cenozoic Eras, based in large part on her personal experiences with both the rocks themselves and the geologists who studied them. According to her sister Louisa, it had been Grace's intention to publish the chapters in book form, but that project was never completed (Milne 1901). If it had been completed, the volume would have probably included a long essay entitled "The Face of the Globe," which was published in three parts in *Good Words*. Subtitled "Short Studies in Physiography," this was a historical account summarizing how, over the previous half century, geologists had come to understand the forces that shape the surface of the earth.

Her essay begins "Amongst subjects of scientific study none is more fascinating, and there are few upon which a greater diversity of opinion still exists, than that part of physical geography and geology relating to the causes which have led to the present configuration of the earth" (Prestwich 1901: 201). After reviewing the debate between Neptunism and Plutonism, she discusses debates over the interior composition of the planet and how mountains are formed. She then summarizes various hypotheses related to ancient glaciers, including work by her famous uncle. Throughout her essay she offers and weighs evidence for and against competing explanations, and demonstrates how various hypotheses were rejected based on the logical application of the methodology of science despite the passions and personal beliefs of scientists. She offers in conclusion, "May it not therefore be a question whether, in another half century, some of the views relating to those great physiographical and geological questions which seem to receive such general acceptance in the present day will not have taken a direction very different from that to which the popular belief is now so strongly guiding them?" before ending with a pious verse penned by Alfred, Lord Tennyson (*In Memoriam A.H.H.*):

> Our little systems have their day;
> They have their day and cease to be:
> They are but broken lights of Thee,
> And Thou, O Lord, art more than they. (qtd. in Prestwich 1901: 228)

Lady Prestwich also wrote popularized accounts of the content of at least two of her husband's technical papers. Her 1879 article "Lochaber and the Parallel Roads" in *Good Words* summarizes her personal experiences visiting these impressive glacial structures with her husband in 1878. After describing their intriguing appearance, and some of their historical, philological, and mythological connotations, she turns to the science, and summarizes three alternate hypotheses on their formation, which she terms the "Fresh-water Lake Theory, with detrital barrier," "Marine theory, with submergence," and "Fresh-water Lake theory, with ice barrier" (Prestwich 1901: 195). All three of these hypotheses are explored at length in her husband's technical article "On the Origin of the Parallel Roads of Lochaber, and Their Bearing on Other Phenomena of the Glacial Period" in the *Proceedings of the Royal Society of London* (Prestwich 1879).

Similarly her 1881 article in *Good Words* entitled "Channel Tunnels and Channel Bridges" first reviews previous plans for building bridges and tunnels across the English Channel. After this historical survey, she describes in detail the various strata that are found in the Channel floor as well as the resulting pros and cons for building a proposed tunnel through each layer. The article is accompanied by a geological map of the area and a geological section, both adapted from her husband's paper on the same topic, *On the Geological Conditions Affecting the Construction of a Tunnel between England and France*, originally presented to the Institution of Civil Engineers (Prestwich 1874). She references her husband's work and paper in the article and describes his conclusion that "these old Paleozoic rocks present most favourable conditions for the piercing of a tunnel, and offer—so far as safety from water is concerned—the surest ground through which to run direct communication between the two countries" (Prestwich 1901: 245). According to her sister (Milne 1901), some of the illustrations in Joseph Prestwich's textbook *Geology: Chemical, Physical, and Stratigraphical* (published in two volumes; 1886 and 1888) were done by Grace, but Mather and Campbell (2007) were unable to find any acknowledgements or citations to Grace within the text. In addition, between 1888 and 1895 she published several articles in *Blackwood's Magazine* and *Leisure House* based on her earlier travels with her uncle (including an account of their meeting with Mary Somerville).

After her husband died on June 23, 1896, 6 months after being knighted, she lived with her sisters while writing a biography of her husband, but spent much of that time as an invalid suffering from heart disease (Mather and Campbell 2007). When she died on August 31, 1899, her passing was noted in the anniversary address by the President of the Geological Society of London, William Whitaker, a singular honor for a woman at that time (Whitaker 1900). In the life of Lady Prestwich we not only bear witness to the power of familial mentorship in providing opportunities for women in science, but the talents of an intelligent amateur scientist and effective science popularizer. In both these traits she echoes the legacy of Mary Somerville, despite the fact that Prestwich's story has thus far escaped the attention of many historians of science.

References

Bakewell R (1819) An introduction to mineralogy. Longman, Hurst, Rees, et al., London

Black A, Black C (1907) Who's who 1907. London

Bowler PJ (2009) Science for all: the popularization of science in early twentieth-century Britain. Chicago UP, Chicago

Buckley A (1876) Short history of natural science. Appleton, New York

Buckley A (1880) The fairy-land of science. Stanford, London

Cook ET, Wedderburn A (1905) The complete works of John Ruskin, vol 18. Allen, London

Daw G (2014) 'On the wings of imagination': Agnes Giberne and women as the storytellers of Victorian astronomy. The Victorian 2(1):1–23

Forsyth JS (1827) The first lines of botany. Bulcock, London

Gates BT (1998) Kindred nature: Victorian and Edwardian women embrace the living world. Chicago UP, Chicago

Giberne A (1879) Sun, moon, and stars: a book for beginners. Carter, New York

Giberne A (1882) The world's foundations or geology for beginners. Carter, New York

Giberne A (1885) Among the stars. American Tract Society, New York

Godding DW (1847) First lessons in geology. Parsons, Hartford

Larsen K (2006) Shooting stars: women directors of the British Astronomical Association meteor section. The Antiquarian Astronomer no 1. pp 75–82
Lightman B (2009) Victorian popularizers of science: designing nature for new audiences. Chicago UP, Chicago
Mather JD, Campbell I (2007) Grace Anne Milne (Lady Prestwich): more than an amanuensis? In: Burek CV, Higgs B (eds) The role of women in the history of geology. Geological Society, London, pp 251–264
Milne LE (1901) Biographical essay. In: Prestwich GA (ed) Essays descriptive and biographical. Blackwood, Edinburgh, pp 1–70
Myers G (1989) Science for women and children: the dialogue of popular science in the nineteenth century. In: Christie J, Shuttleworth S (eds) Nature transfigured: science and literature, 1700–1900. Manchester UP, Manchester, pp 171–200
Osborn HF (1926) A contemporary of Charles Darwin. Science 64:623
Prestwich J (1874) On the geological conditions affecting the construction of a tunnel between England and France. Clowes, London
Prestwich J (1879) On the origin of the parallel roads of Lochaber, and their bearing on other phenomena of the glacial period. Proc R Soc Lond 29:6–21
Prestwich GA (1901) Essays descriptive and biographical. Blackwood, Edinburgh
Ruskin R (1908) The ethics of the dust. Dent and Sons, London
Shteir AB (1996) Cultivating women, cultivating science. Johns Hopkins UP, Baltimore
Shteir AB (1997) Elegant recreations? Configuring science writing for women. In: Lightman B (ed) Victorian science in context. Chicago UP, Chicago, pp 236–255
The American Naturalist (1884) Buckley's winners in life's race. 18(1):47–50
The Atheneum (1887) Animals from life. 3134:690
The Times (1939) Miss Agnes Giberne: a pioneer of popular science. Aug 22:12
Whitaker W (1900) Anniversary address of the president for 1899. Q J Geol Soc 56:i–xcvi

Chapter 11
Conclusion

In 1799 familiar format writer Priscilla Wakefield gratefully observed of her era that "Persons of genius have not thought it unworthy of their talents to compose books purposely for the instruction of the infant mind" (121). `However, as Charles Morgan warned in his largely negative review of Rosina Zornlin's *What is a Voltaic Battery?*, "It is not sufficient to possess knowledge in order to communicate it," because to "write successfully for undeveloped intellects, requires a combination of the highest powers" in order to communicate it (1842: 833). But as we have seen, the works of these women were not only read by *infant minds* and *undeveloped intellects,* but adult women and men as well. Some works were specifically written as textbooks for female seminaries, such as Almira Phelps' *Familiar Lectures* series, or were utilized as textbooks, regardless of the author's initial intent, such as Jane Marcet's *Conversations on Chemistry*. Lady Grace Ann Prestwich and Mary Somerville summarized, digested, and translated scientific papers into a more common, but still very much technically correct, vernacular. As William A. Hamor and Lawrence W. Bass explain in the august professional journal *Science*, "A good translation of a scientific paper is a work of art. It is at once elevated and popular; it manifests to all that which is recondite" (1929: 634). It not only utilizes poetry, and motivates the writing of poetry, it *is* poetry in prose form.

Peter Bowler (2009) reflects that, while the precise timeline associated with the transition of the term *popular* from a compliment into a derogatory connotation is debated by historians, this inevitability was already set into motion by the late nineteenth century. This is certainly apparent in a review of Agnes Clerke's *A Popular History of Astronomy in the Nineteenth Century*, a work somewhat boldly written before the end of said century (1885). In his enthusiastic review, Cambridge astronomer Robert S. Ball notes

> The authoress (for this learned volume is indeed the product of a lady's pen) has modestly described her "History of Astronomy" as a *popular* work. We certainly hope that the book will be as popular as it deserves, and that it will be widely and extensively read. We think, however, that few men of science who use this book will think that it ought to be classed as

K. Larsen, *The Women Who Popularized Geology in the 19th Century*,
DOI 10.1007/978-3-319-64952-8_11

> a popular work in the ordinary acceptation. It might be more correctly described as a masterly exposition of the results of modern astronomy. (1886: 313; emphasis original)

In his review of the same work, William W. Payne, astronomer and publisher of the popular level magazine *The Sidereal Messenger*, reflects that "the years are few since the time that a woman either could or would have written a book on astronomy for popular reading in Europe or America, because the *savants* would have thought such an undertaking strangely unorthodox according to any scientific creed" (1886: 159; emphasis original). Given the success of Mary Somerville's *Physical Geography* and other works in the geological sciences, Payne would likely have agreed that this was also true in geology as well.

Like Mary Somerville, Agnes Clerke (1842–1907) had been largely self-taught in science. Born on February 10, 1842, at Skibbereen, Ireland, the younger daughter of bank manager (later court clerk) John William Clerke and his wife Catherine Mary Deasy, Agnes explained "in the scientific sense I had no education, but just picked up what came my way. I had a childish passion for natural knowledge and read the books accessible to me in my father's library, taught myself with the minimum of assistance some elementary mathematics, and that was all" (Dent 1901: 82).[1] But this is a modest assessment. Like Mary Somerville, Almira Hart Lincoln Phelps, Jane Kilby Welsh, Jane Marcet, Lady Grace Ann Prestwich, Charlotte Murchison, and many other women writers described in these pages, Agnes Clerke continued to educate herself well after her childhood years. She attended public lectures at the Royal Institution and discussed science with some of the most notable professional and amateur astronomers of her day (Huggins 1907). And like the other science writers discussed in this volume Clerke turned his self-education into a mission to educate and inform others.

As we have repeatedly seen in this investigation to uncover the lives and legacies of these women writers, while the popularization of science has been undervalued (and still remains so today), it is nonetheless a vital part of the scientific endeavor. For as trailblazing astronomer Margaret Lindsay Huggins wrote of such work,

> The mission of these special workers is to collect, collate, and digest the mass of observations and papers; to chronicle, in short, on one hand, and, on the other, to discuss and suggest, and to expound; that is, to prepare material for experts, to inform and interpret the general public. There is urgent need of a better-educated public opinion in this country. (1907: 227)

More than a century later, these words still ring true.

Speaking specifically of her close friend, Lady Huggins offered "That such a mission may be a splendid and fruitful one has been shown by Agnes Clerke; what careful preparation it requires, and how much it demands of those who would enter upon it, her career also shows" (Huggins 1907: 227). For despite her fame and commercial success, Agnes Clerke was also not immune to the longstanding prejudices of the men of science who sought to keep a mere popularizer (and specifically a woman) out of the ivory tower of pure science. In his review of Clerke's *Problems*

[1] For more information on Agnes Clerke, see Brück (2002).

in Astrophysics (1903), a survey and synthesis of the current status of the field as she saw it, *Nature* assistant editor R.A. Gregory (who had perennially criticized her work) condescendingly informed Clerke of her proper place (eerily reminiscent of similar complaints about Alice Bodington a decade earlier):

> Her function as an historian is to assimilate and describe, and when she is exercising her talents in this direction she is at her best. She surveys the work from the point of view of a spectator, and should describe fairly and clearly what she sees, without irritating the men who are doing the work by expressing her opinion upon it or suggesting what course they ought to take next. In other words, she should remember that "Passengers are respectfully requested not to speak to the man at the wheel." (1903: 339)

Agnes Clerke not only subsequently brought forth a second edition of an earlier book, but another original popularized (and commercially popular) work, *Modern Cosmologies*, in 1905 (a survey of the formation of the earth and solar system). This was a work that even R. A. Gregory had to grudgingly recommend for the popular level reader (Brück 2002). Jane Kilby Welsh had not been quite so bold in answering her anonymous critics, eventually silencing her scientific pen (but not after publishing the second volume of her *Familiar Lessons in Mineralogy and Geology*). But the quiet inclusion in her will of a bequest to recognize the men of science who had supported the publication of her work speaks volumes in her absence.

As Welsh described (through the voice of one of her fictional characters) in that aforementioned second volume, geology is

> a grand and capricious study; it leads us through the wildest scenes of nature – over the highest eminences – into caverns and grottoes – through lakes and seas. It carries us back to the creation of the globe, and through its history of deluges, earthquakes, volcanoes, hurricanes, and the various other devastations of which its surface affords us undoubted evidence. Indeed, it includes every subject connected with our planet. (1833: 15)

In the twenty-first century geology also truly includes *space* for everyone to take part in this grand adventure to understand not only our world, but the myriad others now known beyond our solar system. The journey to achieve equal access for everyone in science has been rocky, to say the least, and in terms of valuing the popularization of science there still a long journey ahead. But in a reversal of Charles Lyell's famous statement concerning geological processes, it is not possible to understand the present state of the popularization of science without understanding its past. And one now slightly more illuminated piece of that puzzle is the lives and work of some of the women who popularized geology, those who heeded the cartoonish banner displayed on the cover of Delia Woodruff Godding's perhaps ironically named *First Lessons in Geology* and took its single word to heart: Persevere.

References

Ball RS (1886) Astronomy during the nineteenth century. Nature 33:313–314
Bowler PJ (2009) Science for all: the popularization of science in early twentieth-century Britain. Chicago UP, Chicago

Brück MT (2002) Agnes Mary Clerke and the rise of astrophysics. Cambridge UP, Cambridge
Dent EA (1901) Agnes Mary Clerke. J RASC 1(2):81–84
Gregory RA (1903) The spectroscope in astronomy. Nature 68:338–341
Hamor WA, Bass LW (1929) The popularization of science. Science 70:632–634
Huggins ML (1907) Agnes Mary Clerke. ApJ 25:226–230
Morgan C (1842) What is a voltaic battery? by Rosina M. Zornlin. The Athenaeum no. 778. pp 833–834
Payne WW (1886) Book notes: A popular history of astronomy during the nineteenth century. The Sidereal Messenger 5:158–159
Wakefield P (1799) Mental improvement, or the beauties and wonders of nature and art, in a series of instructive conversations, 1st Amer. edn. Greene, New Bedford
Welsh JK (1833) Familiar lessons in mineralogy and geology, vol 2. Clapp and Hull, Boston

Index

K. Larsen, *The Women Who Popularized Geology in the 19th Century*,
DOI 10.1007/978-3-319-64952-8

MIX
Papier aus verantwortungsvollen Quellen
Paper from responsible sources
FSC® C105338

If you have any concerns about our products,
you can contact us on
ProductSafety@springernature.com

In case Publisher is established outside the EU,
the EU authorized representative is:
Springer Nature Customer Service Center GmbH
Europaplatz 3, 69115 Heidelberg, Germany

Printed by Libri Plureos GmbH
in Hamburg, Germany